Engineering Chemistry-II
(For B.E./B.Tech. Students)

Dr. C. Paul Raj

Assistant Professor

University College of Engineering Villupuram

(Constituent College of Anna University of Technology)

Chennai.

Published by

Engineering Chemistry-II

ISBN 978-93-84743-36-9

Author

Dr. C. Paul Raj

Bonfring

309, 2nd Floor, 5th Street Extension, Gandhipuram,

Coimbatore-641 012.

Tamilnadu, India.

E-mail: info@bonfring.org

Website: www.bonfring.org

Phone: 0422 4213231

Preface

The inspiration to textbook to Engineering students comes from the fact that there is a need to learn fundamental concepts in lucid and simple way. It is important to understand the intrinsic facts around each concept which makes the their further studies pleasant and comfortable. The myth that fundamentals of Chemistry is always hard in the point of students should be completely removed and for such purpose it is inevitable to write simple understanding textbooks in a focused manner for specific use.

The guiding scale and principles are simplicity, adequate content under each concept in addition to applied knowledge. The main aim is covering entire syllabus in a single textbook which avoids searching multiple books and taking elaborate notes. It is also kept in mind that some expertise knowledge to improve the forward thinking is essential for the future application of concept in their chosen area of study.

It is also a fact that engineering team of students are heterogeneous and also with different medium of school learning which makes them a Language become a barrier of learning. To overcome this fear simple language is used to explain the c concepts as simple as possible without altering its factual understanding.

From the point of view of author the book has sufficient information and relevant materials for the students to get better score in the examination.

Dr. C. Paul Raj

UNIT I

ELECTROCHEMISTRY

Electrochemical cells – reversible and irreversible cells – EMF – measurement of emf – Single electrode potential – Nernst equation (problem) – reference electrodes –Standard Hydrogen electrode - Calomel electrode – Ion selective electrode – glass electrode and measurement of pH – electrochemical series – significance – potentiometric titrations (redox - Fe^{2+} vs dichromate and precipitation – Ag+ vs CI· titrations) and conductometric titrations (acid-base – HCl vs, NaOH) titrations.

Electrochemistry

Electrochemistry deals with the generation and utilization of electrical energy using chemical reactions.

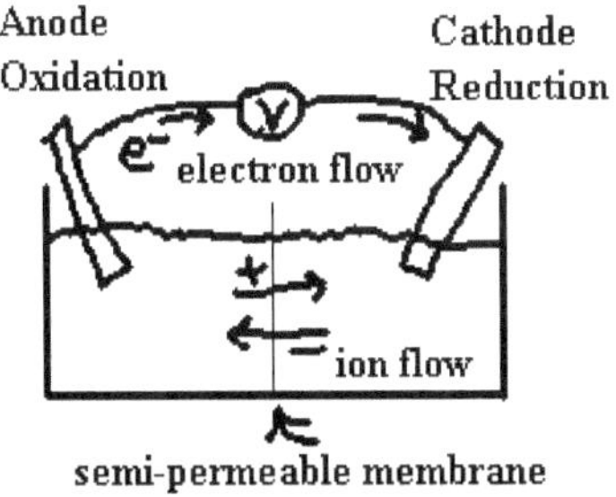

It deals with electron transfer reactions and its applications. Thus electrochemistry is the study of interchange of chemical and electrical energy.

Sl. No	Anode	Cathode
1	The electrode at which oxidation occurs	The electrode at which reduction occurs.
2	It has a negative sign	It has a positive sign.
3	The electrons to be transferred, flow or move from the anode	It accepts electrons.
4	Corrosion occurs at anode	No corrosion at cathode.
5	Positive ions move from anode to cathode.	Negative ions move from cathode to anode.

Sl. No	Oxidation Half Cell	Reduction Half Cell
1	Electrode at which electrons are generated is called oxidation half cell.	Electrode at which electrons are used is called reduction half cell.
2	More electronegative (less electropositive) element act as an oxidation half cell.	Less electronegative (more electropositive) element act as an oxidation half cell.
3	$Zn \longrightarrow Zn^{2+} + 2e^-$	$Cu^{2+} + 2e \longrightarrow Cu$

Cell Reaction

- Over all electrochemical reaction is a cell reaction.
- It is a balanced equation including charges.
- It consists of oxidation and reduction half reactions.

$$Zn + Cu^{2+} \longrightarrow Zn^{2+} + Cu$$

Salt Bridge

It is a device used to give contact between electrodes in a cell. It consists of a glass tube filled with agar and electrolytes like KCl solution. It reduces the liquid junction potential.

Electrochemical Cells

A cell is a device consists of two electrodes dipped into and electrolyte with an external connection for electron flow.

An electrochemical cell is a device for the conversion of electrical energy into chemical energy and vice-versa.

An electrode is a surface on which electron transfer occurs. It also acts as a medium to flow of electrons. An electrode can act as a source of electrons or as a sink for electrons.

Electrochemical cells can be divided into two types, namely Galvanic cells and electrolytic cells.

- The chemical reaction is spontaneous and produces electricity. This is called a voltaic cell or a galvanic cell.
- The chemical reaction is non-spontaneous and is forced by electricity from an external source. This kind of cell is called an electrolytic cell.

a. Galvanic Cells

- In Galvanic cells chemical energy is converted into electrical energy.
- In Gavanic cells spontaneous chemical reaction occurs.
- The free energy change is negative (ΔG = - ve). e. g., Danial cell.

Cell Representation

$$Zn/Zn^{2+}//Cu^{2+}/Cu$$

At anode

$$Zn(s) \longrightarrow Zn^{2+}(aq) + 2e^-(aq)$$

At cathode

$$Cu^{2+}(aq) + 2e^-(aq) \longrightarrow Cu(s)$$

Cell reaction

$$Zn(s) + Cu^{2+}(aq) \longrightarrow Zn^{2+}(aq) + Cu(s)$$

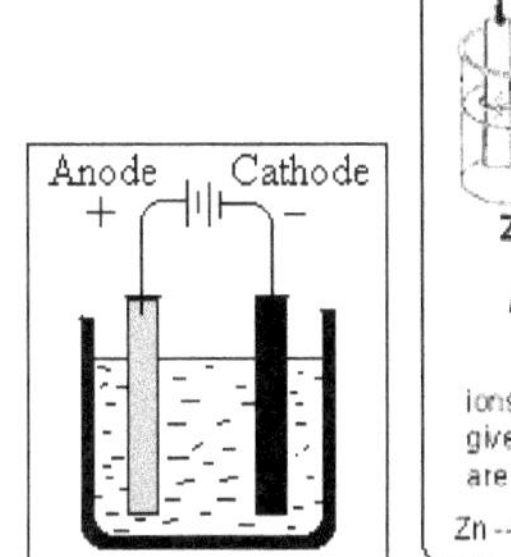

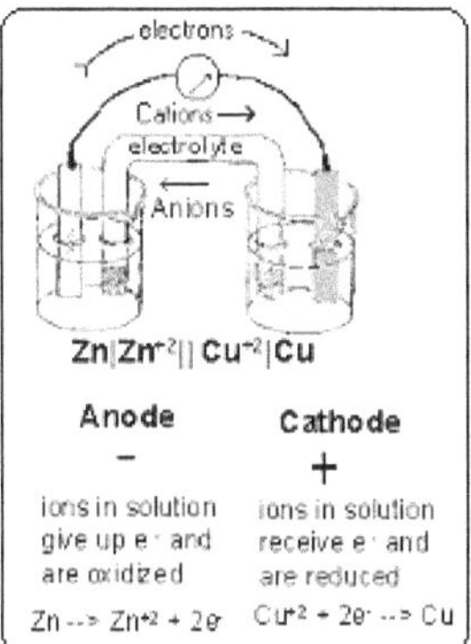

The difference in potential between the electrode reactions is the origin of electromotive force. It is abbreviated as emf and its unit is volts.

b. Electrolytic Cells

Electrical energy is used by electrolytic cells and brings a physical or chemical change.

There is no spontaneous reaction in the cells. It is driven by external electrical energy. Thus electrochemical process is non-spontaneous and change in free energy is positive ($\Delta G = + ve$). Storage batteries and electrolysis process in metallurgy are examples for electrolytic cells.

Difference Between Electrochemical Cells and Electrolytic Cells

Sl.No	Electrochemical Cell	Electrolytic Cell
1	Chemical reaction produces electrical energy.	Electrical energy drives a chemical reaction.
2	Electrical energy is produced.	Electrical energy is consumed.
3	Electrochemical cell is formed.	A process called electrolysis occurs.
4	Nernst equation governs a electrochemical process.	Faraday law of electrolysis is applied for quantitative analysis.
5	Anode is negatively charged. Cathode is positively charged.	Anode is positively charged. Cathode is negatively charged.
6	e.g., Galvanic cell	e.g., electrolysis of water

Reversible and Irreversible Cells

Sl.No.	Reversible Cells	Irreversible Cells
1	An external emf supplied exactly to the emf of the cell no cell reaction occurs.	Does not satisfy the condition.
2	An external emf is infinitesimally smaller than the cell's emf cells reaction occurs and current flows to that extent.	Need not follow the same.
3	An external emf is infinitesimally greater than the cell's emf cells reaction occurs in the opposite direction and current flows to that extent in the reverse direction.	Need not satisfy such condition
4	Eg. Danial cell	$Zn(s)/dil.H_2SO_4/Cu$

Irreversibility of $Zn(s)/dil.H_2SO_4/Cu$ Cell

Condition	Anode Reaction	Cathode Reaction	Cell Reaction
No external emf	$Zn \longrightarrow Zn^{2+} + 2e^-$	$2H^+ + 2e^- \longrightarrow H_2$	$Zn + 2H^+ \longrightarrow Zn^{2+} + H_2$
Infinitesimally greater emf	$Cu \longrightarrow Cu^{2+} + 2e^-$	$2H^+ + 2e^- \longrightarrow H_2$	$Cu + 2H^+ \longrightarrow Cu^{2+} + H_2$

Since hydrogen gas is evolved there is no possibility of reversing the cell reaction by external emf. Rather when applied an external emf greater than the cells emf cell reaction is different but not the reverse of the forward reaction. Thus it is an irreversible cell.

Electromotive Force (EMF)

The difference in potential between electrodes in a cell, which causes the flow of electrons, is electromotive force.

It depends on the relative tendency of the electrodes to loss or gain electrons when they are dipped in an electrolyte. It is also the measure of redox characteristics of the electrodes in a particular electrolytic solution.

Origin of EMF

The origin of EMF is a flow of current (electrons) from an electrode of higher oxidation potential to another electrode of lower oxidation potential.

EMF varies with temperature and concentration of the electrolyte in accordance with Nernst equation. It also varies with changing electrode partner. EMF is expressed in volts. The value of emf is either zero (no cell reaction) or a positive (forward reaction) of negative (backward reaction) values.

Measurement of EMF

Poggendorff's Compensation Principle

The principle states that, *"if emf of unknown and known cells are equal and opposite direction, there will not be any current flow in the galvanometer."*

Instrument

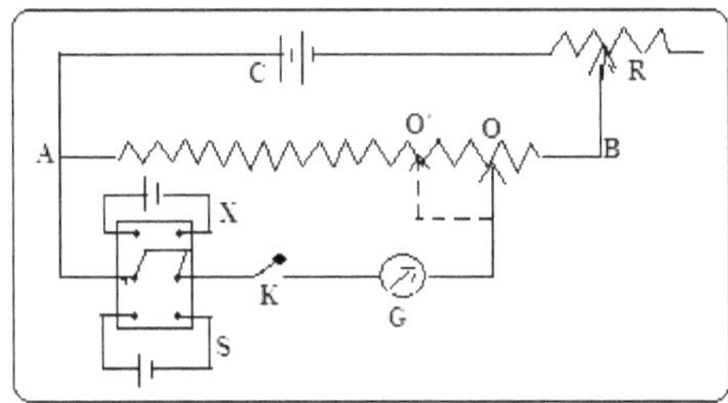

EMF is measured using potentiometer. A cell (C) having emf greater than the one to be measured is connected through a resistance (R). A galvanometer (G) is connected to a sliding wire of uniform thickness. A double pole-double through (DPDT) connects a standard cell (S) or cell having unknown emf (X) with G on one side and AB at the other side.

Experiment

- Using DPDT a connection is established with the battery C and the standard cell S.
- Rheostat is adjusted by moving along the sliding wire AB so that no current in G.
- The null point O' is noted and measured the distance AO'.
- The experiment is repeated for unknown cell (X). The null point O is identified and measured the distance AO.

Calculation

$$\frac{\text{Unknown EMF}}{\text{Known EMF}} = \frac{\text{AO}}{\text{AO}'} \quad \text{Unknown EMF} = \frac{\text{AO}}{\text{AO}'} \times \text{Known EMF}$$

The standard cell must be reliable and should give uniform emf for accurate measurement. Usually Weston-cadmium cell is used as a standard cell. It consists of mercury-mercurous sulphate as positive electrode and 12.5% cadmium amalgam as negative electrode. The emf of Watson cell is 1.018 V at 298 K with temperature coefficient of 0.00005 VK^{-1}. Potentiometer requires small current and thus suitable for emf measurement. On the other hand voltmeter draws appreciable of amount of current from the cell and is unsuitable for emf measurement.

Single Electrode Potential

The tendency of an electrode to lose or gain electrons when it is in contact with its own ion in solution is called single electrode potential.

If the electrode has the tendency to accept electrons it is termed as reduction and thus the potential developed is reduction potential.

$$Cu^{2+} + 2e^- \longrightarrow Cu \quad \text{(reduction potential)}$$

In contrast if the electrode has the tendency to loss electrons it is termed as oxidation and thus the potential developed is oxidation potential.

$$Zn \longrightarrow Zn^{2+} + 2e^- \quad \text{(oxidation potential)}$$

For a particular electrode under given condition if reduction potential is 1.5 V and it oxidation potential is taken as -1.5 V and vice versa.

It is not possible to determine absolute value of single electrode potential. Only relative value can be measured or calculated with respect to a reference.

For this purpose potential of standard hydrogen electrode (hydrogen bubbled at 1 atmosphere pressure with 1 M H^+ ion solution) is taken as zero. With respect to standard hydrogen electrode potential of all the other electrodes are determined. Potential calculated using standard hydrogen electrodes (SHE) are specifically referred as potentials on hydrogen scale.

Since construction and maintenance of standard hydrogen electrode is difficult other reference electrodes like saturated calomel electrode (SCE) is in use.

Sign of Electrode Potential

According to IUPAC convention the electrode undergoes reduction is (consumes electrons) given positive sign when connected to standard hydrogen electrode. A negative sign for the electrode undergoes oxidation.

Further electrode undergoing oxidation is written in the left and electrode undergoing reduction is written in the right in a cell representation.

Standard Electrode Potentials (E^0)

It is the potential developed when the pure metal is in contact with its ions at one molar concentration at a temperature of 25°C or 298K. When a Zn rod of any length is dipped in 1M $ZnSO_4$ solution, standard electrode is formed and the potential developed is called standard zinc electrode potential (E^0_{Zn}). The standard zinc electrode is represented as Zn/Zn^{+2} (1M).

In case of a gas electrode, the standard electrode potential (E^o) is defined as the potential developed at the interface of the gas and solution containing its own ions when an equilibrium is established between the gas at a pressure of 760 mm of Hg and the ions in solution of unit concentration.

When the H_2 gas at a pressure of 1atm is bubbled through HCl of 1 M standard H_2 electrode is formed and the potential developed is called standard hydrogen electrode potential (E^o_{H2}) whose magnitude is considered to be 0.

Electrochemical Series

Reduction Half Reaction	E^o in volt
$F_2(g) + 2e^- \rightarrow 2F^-(aq)$	+2.87
$Cl_2(g) + 2e^- \rightarrow 2Cl^-(aq)$	+1.36
$O_2(g) + 4H^+(aq) + 4e^- \rightarrow 2H_2O(l)$	+1.23
$Br_2(l) + 2e^- \rightarrow 2Br^-(aq)$	+1.09
$Ag^+(aq) + e^- \rightarrow Ag(s)$	+0.80
$Fe^{3+}(aq) + e^- \rightarrow Fe^{2+}(aq)$	+0.77
$I_2(s) + 2e^- \rightarrow 2I^-(aq)$	+0.54
$O_2(g) + 2H_2O(l) + 4e^- \rightarrow 4OH^-(aq)$	+0.40
$Cu^{2+}(aq) + 2e^- \rightarrow Cu(s)$	+0.34
$2H^+(aq) + 2e^- \rightarrow H_2(g)$	0.00
$Pb^{2+}(aq) + 2e^- \rightarrow Pb(s)$	−0.13
$Sn^{2+}(aq) + 2e^- \rightarrow Sn(s)$	−0.14
$Ni^{2+}(aq) + 2e^- \rightarrow Ni(s)$	−0.23
$Co^{2+}(aq) + 2e^- \rightarrow Co(s)$	−0.28
$Fe^{2+}(aq) + 2e^- \rightarrow Fe(s)$	−0.44
$Cr^{3+}(aq) + 3e^- \rightarrow Cr$	−0.74
$Zn^{2+}(aq) + 2e^- \rightarrow Zn(s)$	−0.76
$2H_2O(l) + 2e^- \rightarrow H_2(g) + 2OH^-(aq)$	−0.83
$Mn^{2+}(aq) + 2e^- \rightarrow Mn(s)$	−1.03
$Al^{3+}(aq) + 3e^- \rightarrow Al(s)$	−1.67
$Mg^{2+}(aq) + 2e^- \rightarrow Mg(s)$	−2.34
$Na^+(aq) + e^- \rightarrow Na(s)$	−2.71
$Ca^{2+}(aq) + 2e^- \rightarrow Ca(s)$	−2.87
$K^+(aq) + e^- \rightarrow K(s)$	−2.93
$Li^+(aq) + e^- \rightarrow Li(s)$	−3.02

An arrangement of elements in the order of decreasing reduction potentials with respect to standard hydrogen electrode is known as electrochemical series.

The redox potential is measured with respect to the standard hydrogen electrode and arranged the E^0 values in decreasing order in order to get the electrochemical series.

Salient Features of the Electrochemical Series

- This series is an important tool that helps in predicting many electrochemical reactions.
- In the electrochemical series the elements that are lower in the series get discharged (lose their charge to become neutral) more easily than the ones above them.
- Hydrogen is also included as a reference point in the series.
- The electropositive character and the reducing power of the elements regularly increase downwards.
- The electronegative character and the oxidizing power of the elements regularly increase upwards.
- All metals placed above hydrogen will displace hydrogen from acids while those below it do not.
- Each element in the series will displace any other element below it from its salt solution. For e.g., when magnesium turnings added in copper sulphate solution, copper is replaced by magnesium.

$$Fe + Cu^{2+} \rightarrow Fe^{2+} + Cu \downarrow$$

$$Mg + Cu^{2+} \rightarrow Mg^{2+} + Cu \downarrow$$

- The element gives electron easily will accept electron with great difficulty.

Significance of Electrochemical Series (ECS)

- Elements with positive emf accept e^- and acts as oxidizing agent and vice versa. Fluorine system (F_2/F^-) has the highest emf and hence it possesses highest oxidizing power and this decreases down the series.
  In the ECS, Lithium has the lowest emf. Hence, it has the highest reducing power and this goes on decreasing when move above the series..
- Fluorine system has the highest emf and hence it is most electro negative.
- Li system has the lowest emf and is highly electro positive.
- A metal system occurring above H_2, displaces H_2 from dilute acids,and also liberates H_2 from water.
- Metal system occurring below H_2 does not displace H_2 from dilute acids, water or steam.
- A metal system with lowest emf as negative electrode and vice versa.
- Spontaneity of electrochemical reaction can be predicted. If emf value is positive the reaction is spontaneous and feasible. If it is negative the reaction is not spontaneous.

Electrode	Oxidation reaction	Standard potential (volts)	Nature
$Li \mid Li^+$	$Li \longrightarrow Li^+ + e^-$	+3.040	reducing agents
$K \mid K^+$	$K \longrightarrow K^+ + e^-$	+2.924	
$Ca \mid Ca^{2+}$	$Ca \longrightarrow Ca^{2+} + 2e^-$	+2.870	
$Na \mid Na^+$	$Na \longrightarrow Na^+ + e$	+2.710	
$Al \mid Al^{3+}$	$Al \longrightarrow Al^{3+} + 3e^-$	+1.660	
$Zn \mid Zn^{2+}$	$Zn \longrightarrow Zn^{2+} + 2e^-$	+0.762	
$Fe \mid Fe^{2+}$	$Fe \longrightarrow Fe^{2+} + 2e^-$	+0.441	
$Cd \mid Cd^{2-}$	$Cd \longrightarrow Cd^{3+} + 2e^-$	+0.403	
$Ni \mid Ni^{2+}$	$Ni \longrightarrow Ni^{2+} + 2e^-$	+0.236	
$Sn \mid Sn^{2+}$	$Sn \longrightarrow Sn^{2+} - 2e$	+0.140	
$Pb \mid Pb^{2+}$	$Pb \longrightarrow Pb^{2+} - 2e^-$	+0.126	
$Pt \mid H_2\ H^+$	$H_2 \longrightarrow 2H^+ - 2e$	0.000	
$Cu \mid Cu^{2-}$	$Cu \longrightarrow Cu^{2+} + 2e^-$	-0.337	
$Ag \mid Ag^-$	$Ag(s) \longrightarrow Ag^+ + e^-$	-0.799	oxidising agents
$Hg \mid Hg^-$	$Hg(l) \longrightarrow Hg^{3+} + 2e^-$	-0.521	
$Cl_2\ Cl^-$	$2Cl^- \longrightarrow Cl_2(g) + e^-$	-1.359	

Metal		Reaction with Water		Reaction with Acid		Oxygen
Potassium	K	$2K + 2H_2O \rightarrow 2KOH + H_2$	Displace hydrogen from cold water, producing their hydroxide and hydrogen.	$2K + H_2SO_4 \rightarrow K_2SO_4 - H_2$	These metals react with acids to liberate hydrogen. However the rate of reaction declines down the group from vigorous to steady.	Oxidise in oxygen, the upper elements burn readily the middle ones only when heated, the end ones oxidise slowly, but don't burn, again showing decreasing reactivity.
Calcium	Ca	$Ca + 2H_2O \rightarrow Ca(OH)_2 + H_2$		$Ca + HCl \rightarrow CaCl_2 + H_2$		
Sodium	Na	$2Na + 2H_2O \rightarrow 2NaOH + H_2$		$Na + HCl \rightarrow NaCl + H_2$		
Magnesium	Mg	$Mg + H_2O \rightarrow MgO + H_2$	Displace hydrogen from hot water or steam, the metal itself is hot, and form the metal oxide and hydrogen	$Mg - H_2SO_4 \rightarrow MgSO_4 + H_2$		
Aluminium	Al	$2Al + 3H_2O \rightarrow Al_2O_3 + 3H_2$		$2Al + 6HCl \rightarrow 2AlCl_3 + 3H_2$		
Zinc	Zn	$Zn + H_2O \rightarrow ZnO + H_2$		$Zn + 2HCl \rightarrow ZnCl_2 - H_2$		
Iron	Fe	$Fe + H_2O \rightarrow FeO + H_2$		$Fe + 2HCl \rightarrow FeCl_2 + H_2$		
Tin	Sn	These metals do not react with water.		$Sn + 2HCl \rightarrow SnCl_2 + H_2$		
Lead	Pb			$Pb + 2HCl \rightarrow PbCl_2 + H_2$		
Hydrogen	H			These metals do not react with acids.		
Copper	Cu					
Mercury	Hg					
Silver	Ag	Very unreactive				
Gold	Au					

Notes:
1. Hydrogen is in the series only as a reference point.
2. The electrochemical series differs slightly to that of the activity series (Junior cert.), namely in the position of calcium.
3. A memory aid mnemonic is **P**addy **C**ould **S**till **M**arry **A** **Z**ulu **I**n **T**he **L**ovely **H**onolulu **C**ausing **M**any **S**trange **G**azes.
4. Memorise the table with the groups, which occur at positions 3, 7, 9 and 12.

Limitations of Electrochemical Series

- It is applicable only to dilute solutions. There shall not be any ionic association or inter-ionic interactions.

- It is useful when there is no net loss of current in the system.

- It considers ideal working condition of an electrochemical system.

- According to electrochemical series Cr and Al are expected to have greater affinity towards electron than iron. Practically reverse is true. Iron corrodes faster than Cr and Al.

Theoretical Calculation of EMF of a Cell

$$\text{Emf of a Cell} = \begin{pmatrix} \text{Reduction Potential of} \\ \text{Right Hand Electrode} \end{pmatrix} - \begin{pmatrix} \text{Reduction Potential of} \\ \text{Left Hand Electrode} \end{pmatrix}$$

Problem

Calculate the emf of the following cell using the given data.

$Zn/Zn^{2+}//Cu^{2+}/Cu$

Electrode potential Cu/Cu^{2+} = -0.76V; Zn/Zn^{2+} = 0.34 V.

Calculation

$$\text{Emf of a Cell} = \begin{pmatrix} \text{Reduction Potential of} \\ \text{Right Hand Electrode} \end{pmatrix} - \begin{pmatrix} \text{Reduction Potential of} \\ \text{Left Hand Electrode} \end{pmatrix}$$

Cu/Cu^{2+} = -0.76V (Oxidation Potential)

Cu^{2+}/Cu = 0.76 V

Zn^{2+}/Zn = -0.34 V.

0.76 –(-0.34) = 1.1 V.

Nernst Equation

The general Nernst equation correlates the Gibb's Free Energy ΔG and the EMF of a chemical system known as the galvanic cell. For the reaction

$a\,A + b\,B = c\,C + d\,D$

$$\text{Equilibrium constant (K)} = \frac{[C]^{c}\,[D]^{d}}{[A]^{a}\,[B]^{b}}$$

Thermodynamically, Change in Gibb's free energy

$\Delta G = \Delta G° + R\,T \ln K$

$-nF\,\Delta E = -n\,F\,\Delta E° + R\,T \ln K \qquad [\because \ \Delta G = -nF\Delta E]$

where R, T, K and F are the gas constant (8.314 J mol^{-1} K^{-1}), temperature (in K), reaction quotient, and Faraday constant (96485 Coulomb) respectively.

Dividing by -nF on both side of the equation gives

$$\Delta E = \Delta E^{0} - \frac{RT}{nF} \ln K$$

This is known as the Nernst equation. The equation allows to calculate the cell potential of any galvanic cell for any concentrations.

It is interesting to note the equation is the relationship between equilibrium constant and the Gibb's free energy. When a system is at equilibrium, $\Delta E = 0$, and $K_{eq} = K$. Therefore,

$$\log K_{eq} = \frac{n\Delta E^{\circ}}{0.0592}$$

Thus, the equilibrium constant and ΔE° are related.

For the cell

$$Zn \mid Zn^{2+} \mid\mid H^{+} \mid H_2 \mid Pt$$

the net chemical reaction

$$Zn(s) + 2\ H^{+} = Zn^{2+} + H_2(g)$$

the standard cell potential $\Delta E^{\circ} = 0.763$ V.

$$Zn(s)/ZnSO_4(aq)//H_2SO_4(aq)/H_2(g),Pt(s)$$

If the concentrations of the ions are not 1.0 M, and the H_2 pressure is not 1.0 atm, then the cell potential ΔE may be calculated using the Nernst equation.

with $n = 2$ in this case, because the reaction involves 2 electrons. The activity of the solid Zn is taken as 1. If the H_2 pressure is 1 atm, the term $P(H_2)$ may also be omitted.

$$\Delta E = \Delta E^{\circ} - \frac{0.0592}{n}\log\frac{P(H_2)[Zn^{2+}]}{[H^{+}]^2}$$

Variations of Nernst Equation

$$\Delta E = \Delta E^{0} - \frac{RT}{nF}\ln\frac{[Product]}{[Reactant}$$

$$\Delta E = \Delta E^{0} - \frac{RT}{nF}\ln\frac{[Oxidant]}{[Reductant]}$$

Application of Nernst Equation

- Theoretical calculation of emf.
- Calculation of equilibrium constant.
- Calculation of activity ions.
- Determination of pH of solutions.
- The Nernst equation also indicates that one can build a cell simply by using the same material for both electrodes, but with different concentrations.

Limitation of Nernst Equation

- Applicable only to dilute solutions.
- Its basis is themodymic one and all limitation of thermodynamics also extended to Nernst equation.
- At very low concentrations of the potential determining ions, the potential predicted by Nernst equation tends to ±infinity. It is meaningless. Only the exchange current density becomes very low.
- In concentrated solutions ionic activities are not unity an assumption used   in the Nernst equation.
- But at higher concentrations, the true activities of the ions must be used. This complicates the use of the Nernst equation, since estimation of non-ideal activities of ions generally requires experimental measurements.

Factors Affecting EMF a System

The Nernst equation is written as follows

$$\Delta E = \Delta E^0 - \frac{RT}{nF} \ln \frac{[C][D]}{[A][B]}$$

It indicates emf is a function of

- Standard emf of the system
- Temperature
- Number of electron involved in the electrode reaction
- Concentration of reactants and products.

Problems

1. Calculate the EMF of the cell

$$Zn(s) \mid Zn^{2+} \ (0.024 \ M) \parallel Zn^{2+} \ (2.4 \ M) \mid Zn(s)$$

Solution

$$Zn^{2+} \ (2.4 \ M) + 2 \ e = Zn \quad \text{Reduction}$$

$$Zn = Zn^{2+} \ (0.024 \ M) + 2 \ e \quad \text{Oxidation}$$

$$Zn^{2+} \ (2.4 \ M) = Zn^{2+} \ (0.024 \ M), \ \Delta E^\circ = 0.00 \text{ - - - - - Net reaction}$$

According to Nernst equation

$$\Delta E = 0.00 - \frac{0.0592}{2} \log \frac{(0.024)}{(2.4)}$$

$$= (-0.296)(-2.0) = 0.0592 \text{ V}$$

2. Show that the voltage of an electric cell is unaffected by multiplying the reaction equation by a positive number.

Solution

Assume the following cell

$$\text{Mg} \mid \text{Mg}^{2+} \parallel \text{Ag}^+ \mid \text{Ag}$$

The cell reaction is

$$\text{Mg} + 2\,\text{Ag}^+ = \text{Mg}^{2+} + 2\,\text{Ag}$$

According to Nernst equation

$$\Delta E = \Delta E^0 - \frac{0.0592}{2} \log \frac{\left[\text{Mg}^{2+}\right]}{\left[\text{Ag}^+\right]^2}$$

If multiply the equation of reaction by 2,

$$2\,\text{Mg} + 4\,\text{Ag}^+ = 2\,\text{Mg}^{2+} + 4\,\text{Ag}$$

$$\Delta E = \Delta E^0 - \frac{0.0592}{4} \log \frac{\left[\text{Mg}^{2+}\right]^2}{\left[\text{Ag}^+\right]^4}$$

which can be simplified as

$$\Delta E = \Delta E^0 - \frac{0.0592}{2} \log \frac{\left[\text{Mg}^{2+}\right]}{\left[\text{Ag}^+\right]^2}$$

Thus, the cell potential ΔE is not affected.

3. The standard cell potential ΔE° for the reaction $\text{Fe} + \text{Zn}^{2+} = \text{Zn} + \text{Fe}^{2+}$ is -0.353 V. If a piece of iron is placed in a 1 M Zn^{2+} solution, what is the equilibrium concentration of Fe^{2+}?

Solution

The equilibrium constant K may be calculated using

$$K_{eq} = 10^{(n\,\Delta E^\circ)/0.0592}$$

$$= 10^{-11.93} = 1.2 \times 10^{-12}$$

$$= [\text{Fe}^{2+}]/[\text{Zn}^{2+}].$$

Since $[\text{Zn}^{2+}] = 1$ M, it is evident that

$[\text{Fe}^{2+}] = 1.2\text{E-}12$ M.

4. From the standard cell potentials, calculate the solubility product for the following reaction

$$AgCl = Ag^+ + Cl^-$$

Solution

There are Ag^+ and $AgCl$ involved in the reaction. The standard reduction potential for the system

$$AgCl + e = Ag + Cl^-, E° = 0.2223 \text{ V} - - - -(1)$$

For Ag^+,

$$Ag^+ + e = Ag, E° = 0.799 \text{ V} - - - - - - (2)$$

Subtracting (2) from (1) leads to,

$$AgCl = Ag^+ + Cl^- \ldots \Delta E° = - 0.577$$

Let K_{sp} be the solubility product, and employ the Nernst equation,

$$\log K_{sp} = (-0.577) / (0.0592) = -9.75$$
$$K_{sp} = 10^{-9.75} = 1.8 \times 10^{-10}$$

Reference Electrodes

An absolute standard for the measurement of electrochemical potentials is not available. So there is an inevitable need for a reference electrode.

Emf of an electrode is measured and referred against the reference electrode.

A reference electrode should have a stable and well defined electrochemical potential (at standard conditions).

A good reference electrode is therefore non-polarizable. It means, the potential of such an electrode will remain stable upon passage of a small current. Indicating the impedance of an ideal reference electrode is zero.

1. Primary Reference Electrode

Standard Hydrogen Electrode (SHE)

The equilibrium potential of Standard Hydrogen Electrode (SHE) is defined as being 0 Volt at a $H^+=1$ and $P_{H2}=1$ atm. It is used as a primary reference.

$$Pt/H_2(1atm), H^+ (1 \text{ M})$$

When acts as an anode

$$H_2 \longrightarrow 2H^+ + 2e^-$$

When acts as a cathode

$$2H^+ + 2e^- \longrightarrow H_2$$

Overall reaction

$$H_2 \longrightarrow 2H^+ + 2e^-$$

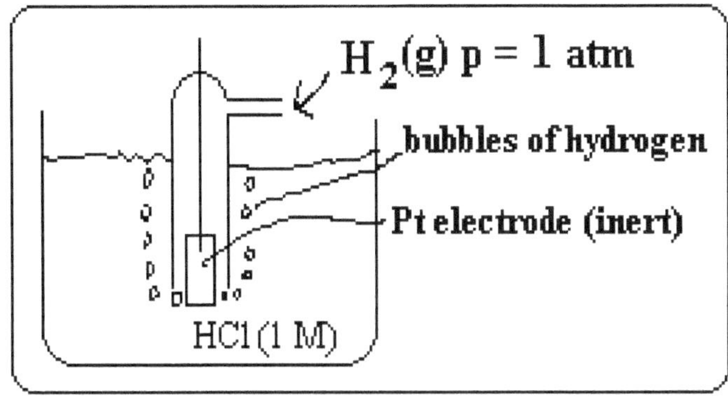

Advantages

- It is reliable and supplies constant emf.
- It is considered as a reference and taken as zero point for emf calculations.
- Small potential is developed on the hydrogen electrode, hence it can be taken as zero.
- In determining the single electrode potential, using SHE as a reference, the potential of the unknown potential will be equal to the emf of the cell.
- The potential developed at standard condition is close to zero.
- SHE is considered as universal electrode. It is the primary reference electrode.

Disadvantages

- Difficult to maintain 1 atmosphere pressure of hydrogen and unit activity of H+ ions.
- It is very sensitive to variation to temperature and pressure.
- Surface poisoning in platinum electrode occurs and thus reproducibility is lost.
- It is not convenient to assemble the apparatus and thus not portable.
- It is difficult to maintain the pressure of hydrogen gas and concentration of HCl.
- It is difficult to get pure, dry hydrogen gas and maintain ideal condition.

2. Secondary Reference Electrodes

Silver/Silver Chloride Reference Electrode

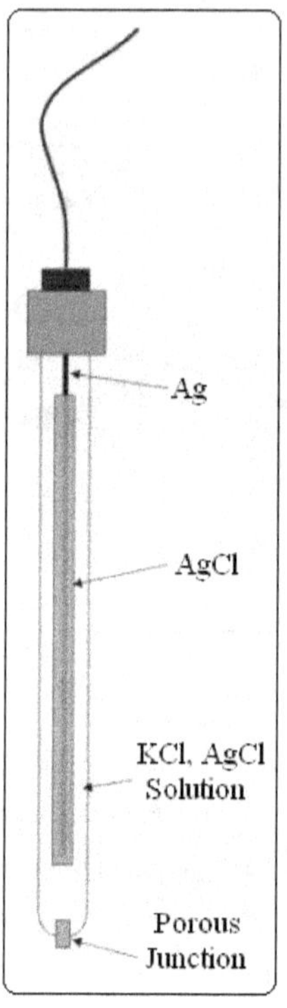

Silver/Silver Chloride Ag/AgCl in saturated KCl constitutes the electrode. This is the most widely used reference electrode, since the use of mercury became less popular. This electrode consists of a Ag wire in contact with AgCl in a saturated KCl solution. This results in an electrode potential of +0.197 Volt vs SHE at 25° C.

Electrodes of this type can be used up to fairly high temperatures (80-100°C). The reference solution is separated from the electrochemical cell by a ceramic frit, or by a glass sleeve (as shown above). Slow leakage of electrolyte takes care of electrical contact.

The silver/silver chloride reference electrode is a widely used reference electrode because it is simple, inexpensive, very stable and non-toxic.

Cell Representation

Ag / AgCl, KCl $_{(Satd)}$

Silver chloride is slightly soluble in strong potassium chloride solutions, so it is sometimes recommended that the potassium chloride be saturated with silver chloride to avoid stripping the silver chloride off the silver wire.

The silver-silver chloride reference electrode develops a potential proportional to the chloride concentration. When the electrode is placed in a saturated potassium chloride solution it develops a potential of 199 mV vs. SHE.

The electrode reaction is:

$$AgCl\ (s) + e^- = Ag\ (s) + Cl^-$$

Advantages

- The silver-silver chloride electrode is the most common due to its ease of manufacture.
- It has wide superior temperature stability, usable even above 130 ^{0}C.

Limitations

- Silver contamination when it is used in pharmaceutical analysis.

Saturated Calomel Electrode (SCE)

Saturated Calomel Electrode (SCE) is Hg/Hg_2Cl_2 in saturated KCl. Traditionally this was the most widely used electrode until the use of mercury was banned from more and more laboratories. The electrode potential is +0.241 V vs. SHE at 25°C.

Hg/ Hg_2Cl_2/KCl(sat)

On reduction

$$Hg_2^{2+}(aq) + 2e^- \longrightarrow 2Hg(l)$$
$$Hg_2Cl_2(s) \longrightarrow Hg_2^{2+}(aq) + 2Cl^-(aq)$$

Oxidation

$$2Hg(l) \longrightarrow Hg_2^{2+}(aq) + 2e^-$$
$$Hg_2^{2+}(aq) + 2Cl^-(aq) \longrightarrow Hg_2Cl_2(s)$$

Electrode Reaction

$$Hg_2Cl_2(s) + 2e^- \longrightarrow 2Hg + 2Cl^-(aq)$$

Concentration	0.1N KCl	1.0 N KCl	Saturated KCl
EMF (V)	0.334	0.280	0.242

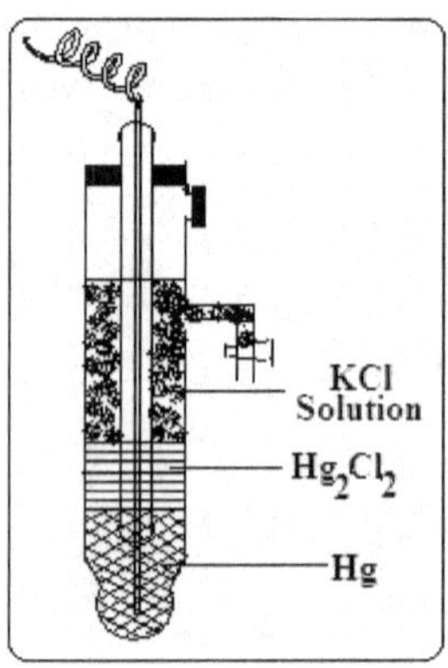

Advantages

- Better stability and reliability as standard electrode.
- Variation with respect to temperature is less.
- Less prone of contamination of electrode since calomel is protected inside.
- It is portable. It is easy to construct and maintain.

Disadvantages

- Mercury toxicity is a prime concern.
- Can not be used at high temperature. This is because above 50°C Hg_2Cl_2 is unstable.

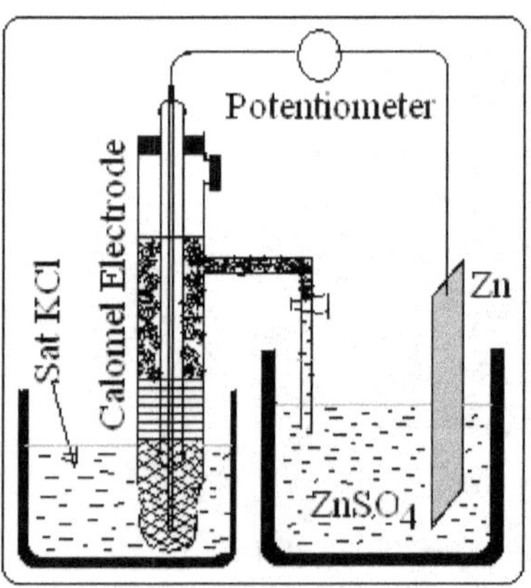

Measurement of Single Electrode Potential

Reference electrode and the electrode whose potential to be determined is connected to a potentiometer. The emf of the cell is noted.

$$\text{Emf of a Cell} = \left(\begin{array}{c}\text{Reduction Potential of}\\ \text{Zn Electrode}\end{array}\right) - \left(\begin{array}{c}\text{Reduction Potential of}\\ \text{Clomel Electrode}\end{array}\right)$$

Emf of the Calomel electrode is known and cell is determined by the experiment, the single electrode potential of the Zn electrode can be calculated.

Glass Electrode

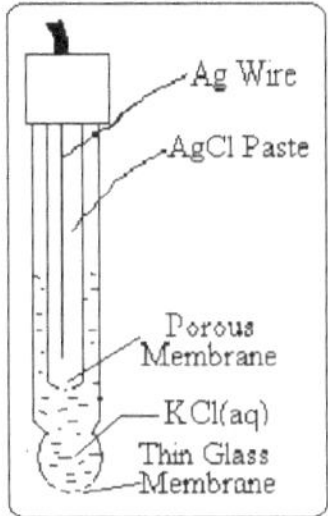

Two solutions of unknown pH values are separated by a glass membrane having high conductivity an exchange of ions takes place across glass. This exchange develops a potential. The magnitude of the potential depends on the difference in pH values of the two solutions.

Pt/glass/H$^+$

Measurement of pH Using Glass Electrode

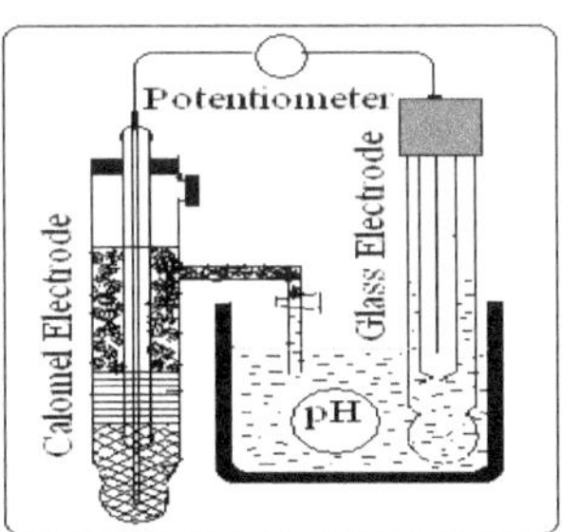

A cell is made using glass electrode and a reference electrode (Calomel electrode). The glass electrode is dipped in a solution for which pH is to be determined. The emf of the cell (E_{cell}) is noted using potentiometer. From Nernst equation,

$$E_{cell} = E°_{ref} + 0.0592 \, V \, (-\log [H^+])$$

$$E_{cell} = E°_{ref} + 0.0592 \, V \times pH.$$

By substituting the values in the above equation pH can be calculated.

Advantages

- Simple instrumentation of pH measurement.
- Strong acid can be used.
- It can give 0.002 pH precision.

Disadvantages

- Needs carefull handling
- Requies regular calibration in sea water buffer
- Strong bases can not be used unless special glass is used

Ion Selective Electrode

Principle

If ions are permeable using semipermeable membrane a concentration gradient is developed across the membrane, which in turn responsible for emf. From the emf developed it is possible to measure the concentration of the ions.

Construction of the Ion-selective Electrodes

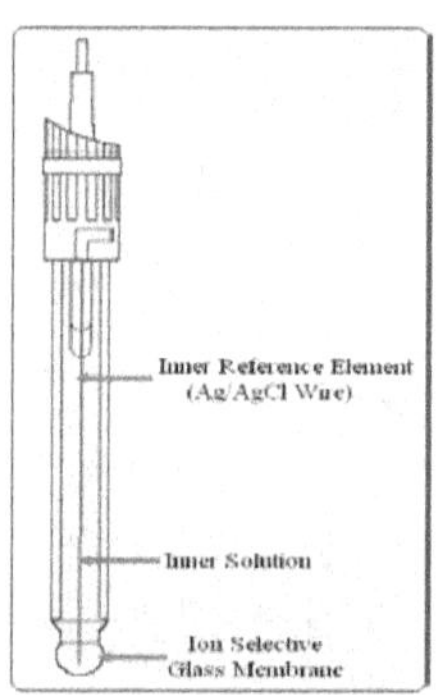

Typical for all kinds of electrodes used in potentiometric ion measurements is an ion-sensitive membrane. This membrane can be prepared as,

- Solid membrane (e.g. glass membrane or crystal membrane)
- Liquid membrane (based on e.g. classical ion-exchanger, neutral or charged carrier)

- Membrane in a special electrode (gas-sensing or enzyme electrode). Typically such a membrane contains an analyte-selective component whichis responsible for the recognition process.

Characteristics of Ion Selective Electrodes

1. Selectivity

The selectivity is one of the most important characteristics of an electrode. It often determines whether a reliable measurement in the sample is possible or not. It indicate the preference of an electrode for the interfering ion.

2. Range of Linear Response

At high and very low target ion activities there are deviations from linearity. Typically, the electrode calibration curve exhibits linear response range between 10^{-1}M and 10^{-5}M.

3. Detection Limit

In practice, detection limit on the order of 10^{-5}-10^{-6}M is measured for most of ion-selective electrodes. The observed detection limit is often governed by the presence of other interfering ions or impurities.

Response time

It was defined as the time between the instant at which the ion-selective electrode and a reference electrode are dipped in the sample solution (or the time at which the ion concentration in a solution is changed on contact with ISE and a reference electrode) and the first instant at which the potential of the cell becomes equal to its steady-state value within 1 [mV] or has reached 90% of the final value.

Advantages of Ion Selective Electrodes

- It is a fast method of analysis.
- Economical to use ion selective electrodes.
- It is possible to do analysis in automated instruments.
- When compared to many other analytical techniques, Ion-Selective Electrodes are relatively inexpensive and simple to use.
- All-solid-state or gel-filled models are very robust and durable.
- They can be in wide range of concentrations in order of magnitude.
- They are invaluable for the continuous monitoring of changes in concentration.

- They are particularly useful in biological/medical applications because they measure the activity of the ion directly, rather than the concentration.
- ISEs are one of the few techniques which can measure both positive and negative ions.
- They are unaffected by sample colour or turbidity.
- ISEs can be used in aqueous solutions over a wide temperature range. Crystal membranes can operate in the range 0°C to 80°C.

Limitations

- The method of analysis is limited to free ions in solution.
- There may be inference from ions which are having closely similar ionic and electrochemical properties. For instance iodide, bromide to the chloride electrode
- It requires careful use, frequent calibration.
- They can achieve accuracy and precision levels of ± 2 or 3% for some ions.
- Specific ion selective electrodes for each type of ion is not always possible.
- It is not always very specific to a particular ion. Thus other ions of close characteristics interfere in its use.

Applications

1. Pollution Monitoring

The presence of ions like CN, F, S, Cl, NO_3 etc., in effluents, and natural waters and drinking water can be monitored.

2. Agriculture

The detection and quantification of NO_3, Cl, NH_4, K, Ca, I, CN in soils, plant material, fertilisers and feedstuffs.

3. Food Processing

- The presence of NO_3, NO_2 in meat preservatives can be detected.
- The quantification of K in fruit juices and wine.
- The study of corrosive effect of NO_3 in canned foods.
- The quantification salt content of meat, fish, dairy products, fruit juices, brewing solutions.
- It is also used in the quantification of Ca in dairy products and beer.]

4. *Paper Manufacture*

The presence of S and Cl in pulping and recovery-cycle of liquors can be determined by ion selective electrodes.

5. *Explosives*

The detection of F, Cl, NO_3 in explosive materials and combustion products can be carried out.

6. *Biomedical Laboratories*

- Detection of Ca, K, Cl ions in body fluids (blood, plasma, serum, sweat).
- Identification of F in skeletal and dental studies.

Types of Ion Selective Electrodes

Polymer Membrane Electrodes

Polymer membrane electrodes consist of various ion-exchange materials incorporated into an inert matrix such as PVC, polyethylene or silicone rubber. After the membrane is formed, it is sealed to the end of a PVC tube. The potential developed at the membrane surface is related to the concentration of the species of interest. Electrodes of this type include potassium, calcium, chloride, fluoroborate, nitrate, perchlorate, potassium, and water hardness.

Solid State Electrodes

Solid state electrodes utilize relatively insoluble inorganic salts in a membrane. Solid state electrodes exist in homogeneous or heterogeneous forms. In both types, potentials are developed at the membrane surface due to the ion-exchange process. Examples include silver/sulphide, lead, copper(II), cyanide, thiocynate, chloride, and fluoride.

Gas Sensing Electrodes

Gas sensing electrodes are available for the measurement of dissolved gas such as ammonia, carbon dioxide, nitrogen oxide, and sulfur dioxide. These electrodes have a gas permeable membrane and an internal buffer solution. Gas molecules diffuse across the membrane and react with a buffer solution, changing the pH of the buffer.

The pH of the buffer solution changes as the gas reacts with it. The change is detected by a combination pH sensor within the housing. Due to their construction, gas sensing electrodes do not require an external reference electrode.

Glass Membrane Electrodes

Glass membrane electrodes are formed by the doping of the silicon dioxide glass matrix with various chemicals. The most common of the glass membrane electrodes is the pH electrode. Glass membrane electrodes are also available for the measurement of sodium ions.

Potentiometric Titrations

Principle

The potential developed by an electrode depends on the concentration of the ions in the electrolyte. As change in concentration of ions affects emf, its measurement can be used for the quantitative determination.

Fe^{2+} Versus Dichromate Titration

Redox titrations can be carried out potentiometrically. The electrode should be reversible to the redox state of the substance under estimation. It is coupled with a suitable reference electrode to form a cell.

On titration of Fe^{2+} ions with dichromate the following redox reaction occurs.

$$Fe^{2+} + Cr_2O_7^{2-} + 14H^+ \longrightarrow 6Fe^{3+} + 2Cr^{3+} + 7H_2O$$

The emf of the Fe^{2+} solution increases slowly on addition of $Cr_2O_7^{2-}$ and a sudden increase at the end point. Based on this observation unknown concentration of Fe^{2+} solution can be determined.

A cell is set up using standard calomel electrode and Fe^{2+}, Fe^{3+} cell.

$$Hg/Hg_2Cl_2(s), KCl\ (sat) \parallel Fe^{3+}, Fe^{2+} / Pt$$

The emf generated during above redox reaction is measured using potentiometer.

Procedure

1. *Calibration of Potentiometer*

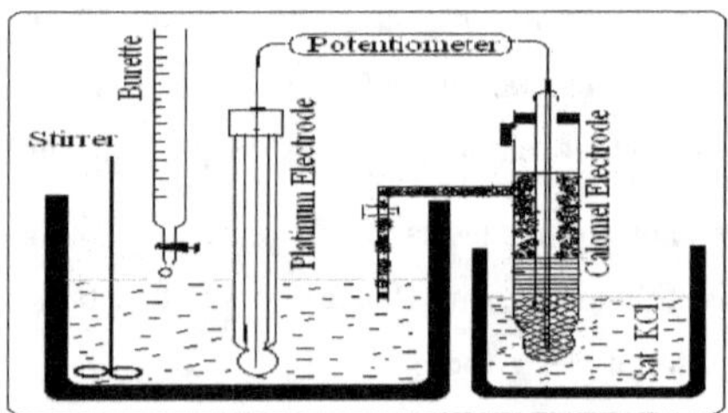

The potentiometer is kept on and connected to the standard cell and read the voltage. The standard value should be 1.018 V, if not adjusted to the value using the calibration knob provided with the equipment.

2. Rough Titration Between Fe^{2+} and $Cr_2O_7^{2-}$

Ex. No.	Vol. of $Cr_2O_7^{2-}$	emf
1	0 ml	
2	1	
3	2	
4	3	
5	4	

In a 100 ml clean conical flask pipette out 20ml of given ferrous ammonium sulphate. The (calomel and platinum) electrodes are dipped into the solution and the emf is noted. Standard potassium dichromate is added from taken burette in aliquots of 1 ml. Every 1 addition of dichromate solution the emf is noted. The values are tabulated.

Initially there is a gradual increase in emf and then a sudden increase near the end point. The volume of dichromate solution near the sudden increase in emf is noted. Based on this value a fair titration is performed.

3. Fair Titration Between Fe^{2+} and $Cr_2O_7^{2-}$

Sl. No	Volume of $K_2Cr_2O_7$	EMF (Volts)	ΔE (Volts)	ΔV (ml)	$\Delta E/\Delta V$	Average Vol. (ml)
1						
2						
3						
4						
5						

In a 100 ml clean conical flask pipette out 20ml of given ferrous ammonium sulphate sample. Dip the (calomel and platinum) electrodes into the solution.

In this titration dichromate from burette is added 1 ml before the end point (as measured from the previous experiment. The emf is noted. Then emf is noted for every 0.5 ml addition of dichromate for two times. After this experiment is continued for every 0.2 ml addition of dichromate till further appreciable increase in emf. The values are tabulated.

A. Graphical Evaluation of Volume of $Cr_2O_7^{2-}$

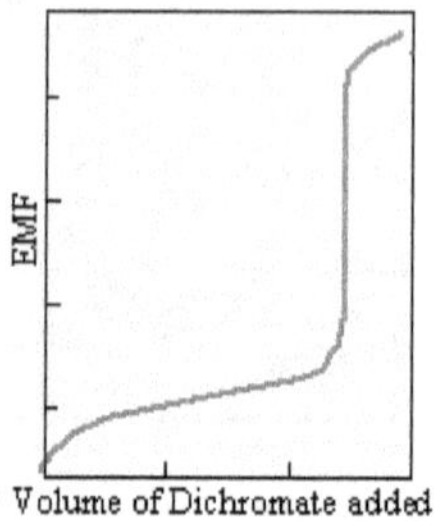

A graph is drawn between volume of $Cr_2O_7^{2-}$ in x axis and emf in y axis. It gives a sigmoid graph. A fair graph is drawn between average volume of $Cr_2O_7^{2-}$ in x axis and $\Delta E/\Delta V$ in y axis. It gives a peak with a baseline. The tip of the peak is taken as an accurate volume of $Cr_2O_7^{2-}$ required to neutralize 20 ml of Fe^{2+} (V_2).

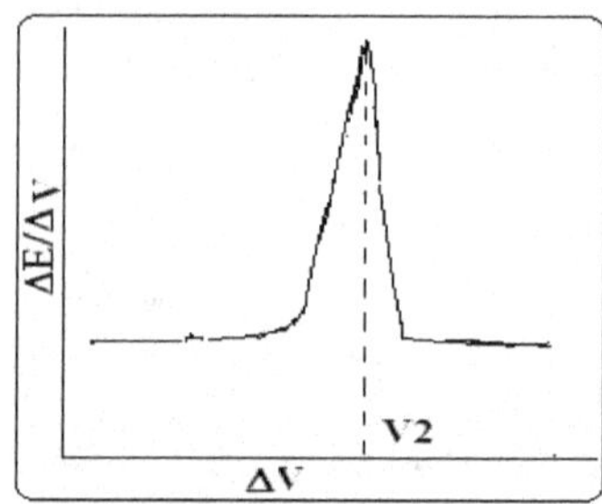

B. Calculation of Normality of Fe^{2+} Solution

Volume of Fe^{2+} solution (V_1) = 20 ml

Strength of Fe^{2+} solution (N_1) = ?

Volume of $Cr_2O_7^{2-}$ solution (V_2) =   ml

Normality of $Cr_2O_7^{2-}$ Solution (N_2) = $V_1N_1 = V_2N_2$

$$N_2 = \frac{V_1 N_1}{V_2}$$

Normality of Fe^{2+} solution = N

C. Calculation of amount of Fe^{2+}

The amount of Fe^{2+} present in the given sample = Normality x Equivalent Weight of ferrous ammonium sulphate (392).

Ag⁺ Versus Cl⁻ Titration

$$Ag^+ + NO_3^- + K^+ + Cl^- \longrightarrow AgCl + K^+ + NO_3^-$$

A cell which is reversible to chloride ion is set up as follows.

Ag/AgCl/Cl-//SCE

According to Nernst equation emf is given by

$$E = E_{SCE} + 0.0591 \log\left[Ag^+\right]$$

The emf decrases slowly first hence the concentration of Ag⁺ decreasing on addition of KCl. As the end point approaches a rapid increase since the concentration of Ag is very small. After end point emf remains almost constant.

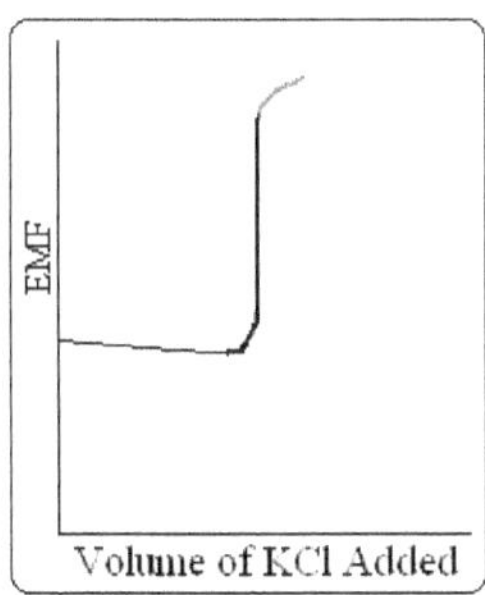

Advantages of Potentiometric Titrations

- Coloured solutions can be used in the titration where as in conventional indicator based titration it can not be done.
- Prior information regarding the strength of solutions need not be required.
- Very dilute solutions can be used.
- No need of indicators and thus the concept of selecting appropriate indicator do not arise.
- No Special care required near end point.

Disadvantages of Potentiometric Titrations

- Selection of appropriate electrode depending upon the use.
- The results are not reliable for gas evolving titrations due to overvoltage.
- Impurities interfere a great extent.
- There should be redox reaction.

Conductometric Titrations

Conductance is the ease with which the ions move in solution. The principle of conductometric titration is electrical conductance depends on the mobility of ions in solution.

Acid-Base (HCl vs NaOH) Titration by Conductometry

HCl and NaOH are strong acid strong bases. They are completely ionized in solution (Ostwald's Law) as follows.

$$H^+_{(aq)} + Cl^-_{(aq)} + Na^+_{(aq)} + OH^-_{(aq)} \longrightarrow Na^+_{(aq)} + Cl^-_{(aq)} + H_2O$$

On initial addition o NaOH conductance decreases due to the formation of unionized water at the expense of H^+ and OH^- ions. Thus conductance decreases gradually.

At end point all H^+ ions of acid is neutralized and thus reaches the minimum. After the end point further addition of base increases the conductance of the solution due the addition of fast moving OH^- ion. The minimum point is the end point of the titration. It is used for the calculation of finding out the unknown concentration.

Titration of Weak Acid against Strong Base

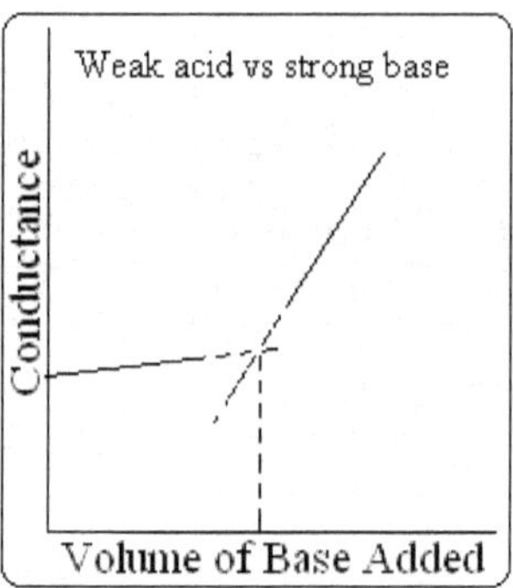

Weak acid is feebly ionized and thus in ionic equilibrium. Due to poor ionization conductance is low.

$$CH_3COOH + Na^+ + OH^- \longrightarrow CH_3COO^- + Na^+ + H_2O$$

On addition of base conductance increases since ionic salt is formed. When all the acid is completely neutralized, further addition of base leads to sudden increase in conductance. This is because of fast moving hydroxyl ions from the base.

By graphical evaluation end point is obtained. It is used for the quantitative estimation.

Titration of Strong Acid against Weak Base

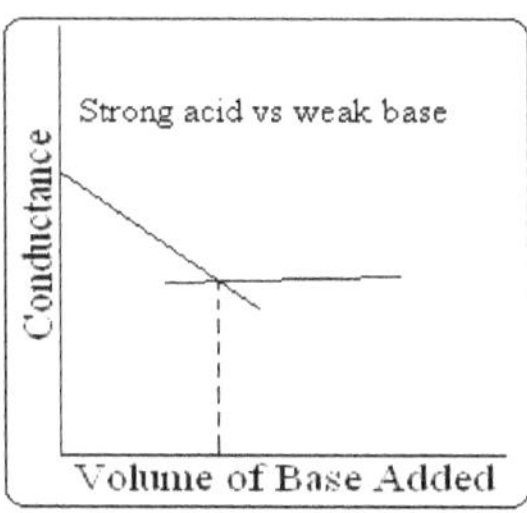

The conductance falls first because of partly fast moving H^+ ions are removed as unionized water and partly by slow moving NH_4^+ ions.

$$H^+ + Cl^- + NH_4OH \longrightarrow NH_4^+ + Cl^- + H_2O$$

After neutralizations excess addition of weakly ionized base (ammonium hydroxide) will not increase the conductance appreciably. End point is calculated by graphical evaluation and is used for quantitative determination.

Titration of Mixture of Acids against Strong Base

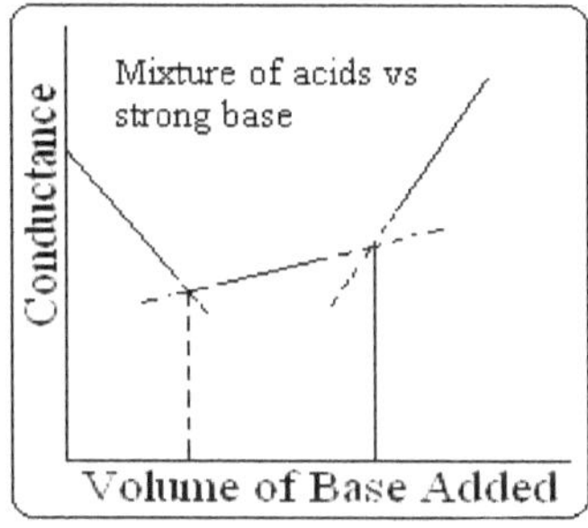

The variation of conductance on titration of mixture of strong acid and weak acid with a strong base is shown in graph. Initially strong acid is neutralized and then weak acid undergo neutralization. From the graph volume required for complete neutralization of strong acid and weak acid can be evaluated. From the values concentration of individual acids in a mixture can be quantified.

Advantages of Conductometric Titrations

- Relatively unfavourable equilibria can be studied.
- Very dilute solution can be titrated.
- No Special care required near end point.

- Coloured solutions can be used.

- Mixture of acid can be titrated.

Disadvantages of Conductometric Titrations

- It has the disadvantage of the lack of sensitivity for surfactants with low concentrations.

- The solution under study has a good conductivity.

- Non-ionic equilibria can not be studied by conductometric titrations.

- Very weak organic acid and base can not be studied conductometrically.

Assignment Questions

1. Distinguish the following (a) cell and battery (b) Electrochemical cell and electrolytic cell.

2. Explain the use of Nernst equation for the determination of equilibrium constant.

3. Explain the estimation o pH using glass electrode and give its draw backs.

4. Explore the application of electrochemistry in medical diagnosis.

5. Weather Ce3+(a=1) ions reduce chlorine to Cl-(a=1) at 298K based on the following data.

$$E^\circ_{Ce4+,3+}/Pt = 1.82\ V; \quad E^\circ_{Cl-,Cl2+}/Pt = 1.3595\ V$$

Self Test

1. Explain the need for standard reference electrodes for the measurement of emf.

2. Copper ions cannot displace H+ ions from acid whereas zinc does the same-explain.

3. From the given values calculate E° value fo MnO_4^-.

    MnO_4^- 0.564 MnO_4^{2-} 2.26 MnO_2 0.95 Mn^{3+} 1.51 Mn^{2+} -1.18 Mn

    Among MnO_4^- and MnO_4^{2-} which is better oxidizing agent?

4. Apply Nernst equation for finding out the spontaneity of a reaction and identify the limitations of Nernst equation.

5. Analyze the need for many standard electrode for the determination of emf.

6. Describe the experimental determination of pH using glass electrode.

7. Explain the superiority of potentiometric titration over titrometry using a suitable example.

8. Bring out the behaviour of acids and bases in conductometric titrations.

9. What are the salient features of ion selective electrodes? Give one electrode which specifically use for detection and quantification of ions.

10. Explain membrane electrode and membrane potential.

CORROSION AND CORROSION CONTROL

Chemical corrosion – Pilling – Bedworth rule – electrochemical corrosion – different types – galvanic corrosion – differential aeration corrosion – factors influencing corrosion – corrosion control – sacrificial anode and impressed cathodic current methods – corrosion inhibitors – protective coatings – paints –constituents and functions – metallic coatings – electroplating (Au) and electroless (Ni) plating.

Corrosion

The spontaneous and slow process of deterioration of metals or alloys by environment is termed as corrosion.

Rusting of iron is the classical example of corrosion that has been encountered in day to day life.

Corrosion is considered as the reverse of metallurgy. In metallurgy metal is extracted from a metallic compound. In corrosion metal goes to compound.

In general it is a surface phenomenon. Corrosion is reverse of metallurgy. The driving force for corrosion is greater stability of metals in its compound form. The high reactivity of atmospheric oxygen and high oxidation potential of metals are the reason for corrosion.

Impact of Corrosion

It has been estimated that economic loss in the world due to corrosion is about 6 billion dollars per year. In India the average loss is about 250 crores per annum. In addition there are cases of life loss and failure of structures, which are planned meticulously. There is no natural cure for corrosion. There is only prevention and control measures to decrease the corrosion.

Consequences of Corrosion

Maintenance Costs

Substantial savings can be obtained in most types of chemical plants through the use of corrosion-resistant materials of construction.

Contamination of Product

Contamination decreases the value of the product. It also causes adverse effect in the manufacture of transparent plastics, food products, and drugs. In the manufacture and transport of hydrogen peroxide, the contamination by metallic impurities causes explosion.

Plant Shutdowns

Far too often plants are shut down or portions of the process stopped because of unexpected corrosion failures.

To increase its production, the temperature of the cooling medium in a heat exchanger system was lowered and the time required per batch decreased.

Stress-corrosion cracking of the vessels occurred quickly, and finally a plant was shut down with production delayed for some time.

Loss of Valuable Chemicals

Particularly during recent years, the products or intermediates handled in a chemical plant are often expensive. There is a substantial production losses of these materials because of corrosion failures.

Safety

In manufacturing concerns handling of chemicals at high temperatures and pressures, explosive materials, and acids and so on. The corrosion failures cause severe injury or loss of life.

Classification of Corrosion

1. Chemical (Dry) Corrosion

The process of corrosion by the chemical action of environmental chemicals on metals is termed as chemical corrosion. It occurs in the absence of moisture or water and thus it is a dry corrosion. Chemical corrosion is due to the presence of gases like O_2, SO_2, CO_2, Cl_2, H_2S, etc. The extent of corrosion is proportional to the chemical reactivity of the metals with these gases.

(a) Hydrogen Embrittlement

The process evolution of hydrogen by gases like H2S in contact with metal is known as hydrogen embrittlement.

The mechanism is formation of atomic hydrogen and combination of it leads to the formation of hydrogen gas.

$$H_2S + Fe \longrightarrow FeS + 2H$$

$$2H \longrightarrow H_2$$

The consequence of hydrogen embrittlement is formation of gas voids, holes, etc inside the metal. It leads to cracks and failure of metal's strength.

(b) Decarburization

Metals at high temperature catalyze thermal decomposition of hydrogen. It gives atomic hydrogen. Iron in steel at high temperature when in contact with hydrogen gives atomic hydrogen. It reacts with carbon present the steel leading to the evolution of methane. It depletes the carbon content in the steel. So it is called as decarburization.

$$H_2 \longrightarrow 2H$$
$$C + 4H \longrightarrow CH_4$$

The process leads to accumulation of gases inside the steel. It develops high pressure at high temperatures and thus develope cracks.

(c) Liquid Metal Corrosion

It is the dissolution of metal in liquid metal. It includes amalgam formation with mercury. It can be at high temperatures in which one of the metals melts and the other metal dissolves in it. It follows the phase equilibrium phenomena.

The consequence is metal loses its mechanical strength. It leads to the accumulation of localized liquid-solid equilibrium phases. It changes in the composition of alloys, etc.

(d) Oxidation Corrosion

It is the direct attack of oxygen on reactive metals. Alakli metals are more prone for oxidation corrosion.

$$M \longrightarrow M^{n+} + ne^-$$
$$O_2 + e^- \longrightarrow O_2^-$$
$$M^{n+} \ nO_2^- \longrightarrow nMO_2$$

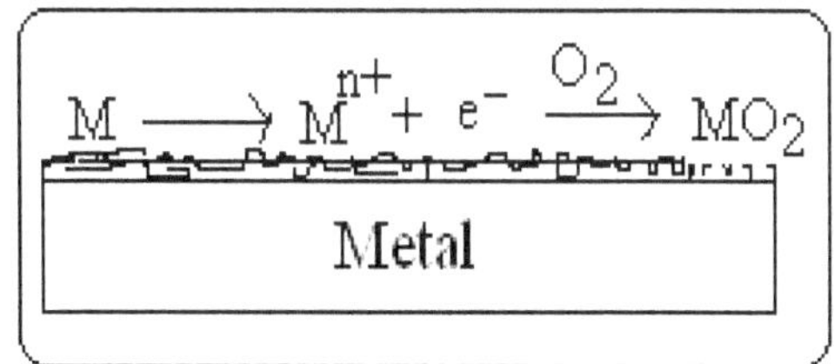

Pilling-Bedworth Rule

Corrosion of metals by oxygen is known as oxidation corrosion. It leads to the formation of metal oxides. Based on the volume of oxide formed the process may be accelerated or prevented which is stated as Pilling and Bedworth rule.

1. When the volume of metal-oxide formed is greater than the atomic volume of metal it is protective oxide.

Aluminum oxide has greater volume than aluminum. So it forms a tight non-porous layer leading to protection from further corrosion. The oxides of Sn, Pb and Cu are also protective oxides.

2. If the volume of metal-oxide formed is less than the atomic volume of metal it is non-protective oxide.

The oxides of alkali (Na, K, etc) and alkali earth metals (Mg, Ca, etc) are porous and non-protective.

3. It the oxide formed is unstable the metal is more stable from corrosion.

The oxides of Pt and Au are unstable and thus decompose back to metal. The Pt and Au are noble metals.

4. Volatile oxides enhance the corrosion. This is because oxide formed soon escapes from the surface leaving fresh surface for corrosion.

Molybdenum oxide (MoO_3) is a volatile oxide and thus Mo very prone for corrosion.

5. The ratio between the volume of oxide and metal atom is known as Pilling Bedworth ratio. If the ratio is more than one it is a protective oxide. It is less than one for non-protective oxides.

Oxides of alkali (Li, Na, K) and alkaline earth metals (Mg, Ba) oxides are having less volume than atomic volume of metals. So they form a porous layer and enhance the further corrosion.

2. *Electrochemical (Wet) Corrosion*

Corrosion accompanied by flow of electrons by electrochemical cell formation is electrochemical corrosion. It requires an electrolytic medium for electron transfer between electrodes. There will be distinct cathodic and anodic reactions.

Evolution of Hydrogen

It occurs in acidic environment. Metal dissolves in the medium with the liberation of electrons. It in turn accepted by hydrogen ions and reduced to hydrogen gas.

$$Fe \longrightarrow Fe^{2+} + 2e^- \text{ (Oxidation)}$$

$$2H^+ + 2e^- \longrightarrow H_2 \text{ (Reduction)}$$

Over all reaction

$$Fe + 2H^+ \longrightarrow Fe^{2+} + H_2$$

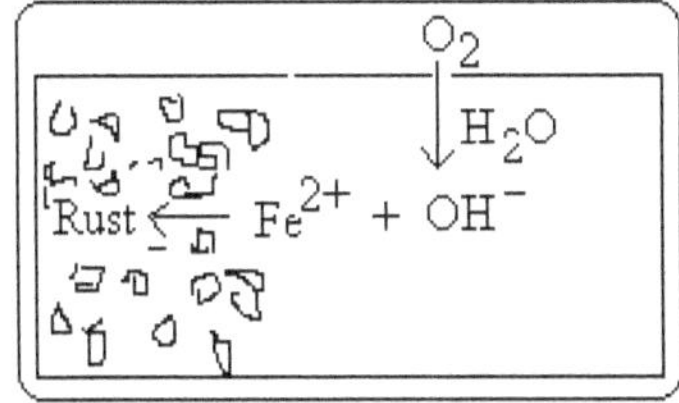

Oxygen Absorption

In this process oxygen is consumed and thus oxide is the product leading to corrosion. Rust is the consequence of the process. Rusting of iron can be explained as follows.

$$Fe \longrightarrow Fe^{2+} + 2e^- \text{ (Oxidation)}$$

$$1/2O_2 + H_2O + 2e^- \longrightarrow 2OH^-$$

$$Fe^{2+} + 2OH^- \longrightarrow Fe(OH)_2$$

Depending upon the availability of oxygen either it gives hydrated yellow ferric oxide ($Fe_2O_3.H_2O$) in excess oxygen or black magnetite, Fe_3O_4 (limited oxygen).

Difference between Dry and Wet Corrosion

Sl. No.	Chemical (Dry) Corrosion	Electrochemical (Wet) Corrosion
1	Occurs in the absence of moisture.	It occurs in the presence of moisture or water.
2	Chemical reaction responsible for corrosion	An electrolytic cell is formed.
3	It leads to the formation of compound(s)	Metal gets dissolved from negative electrode.
4	Pilling-Bedworth rule applies	Need not follow the rule
5	Corrosive chemicals can drive the process.	Need not have influence of such gases.
6	e.g. Rusting of iron in dry air.	e.g., Rusting of ship.

Galvanic Series

The Galvanic Series is a list sorted by corrosion potentials for various alloys and pure metals in sea water.

$$Mg > Be > Al > Cd > Pb > Cu > Ni > Steel > Fe \text{ (Cast)} > Ti > Ag > Au > Pt > Graphite$$

Active (Anode)   Galvanic Series   Decreasing Order of Corrosion   Noble (Cathode)

The emf series is a list of half-cell potentials for standard state conditions measured with respect to the standard hydrogen electrode, while the Galvanic Series is based on corrosion potentials in sea water.

Uses of Galvanic Series

- It related the corrosion tendency in environmental condition.
- Galvanic series relationships are useful as a guide for selecting metals to be joined, while welding, etc.
- It guides in selecting alloying elements.
- It helps the selection of metals having minimal tendency to interact galvanically.

Limitations of Galvanic Series

The series does not provide any information on the rate of galvanic corrosion and thus serves as a basic qualitative guide only.

It is based on only corrosion potential. Thus general chemical reactivity of metals can not be correlated.

Difference between Electrochemical Series and Galvanic Series

Sl.No.	Electrochemical Series	Galvanic Series
1	Arrangement of elements in their increasing order of reduction potential.	Decreasing order of Corrosion of elements in sea water.
2	It is relevant only for metals.	It includes metals and alloys.
3	Position in the series is fixed.	May vary when the element present in alloys.
4	It comprises of metals and non-metals.	It is for metals and alloys.
5	It predicts relative electron transfer potential.	It relates the relative corrosion tendency.
6	Alloys are not included.	Alloys are included

Different Types of Corrosion

   i. Uniform, or general attack

  ii. Galvanic, or two-metal corrosion

 iii. Crevice corrosion

 iv. Differential Aeration Corrosion

  v. Pitting Corrosion

 vi. Intergranular corrosion

 vii. Selective leaching, or parting

viii. Erosion corrosion

 ix. Stress corrosion.

  x. Microbial Corrosion

(i) Uniform Corrosion

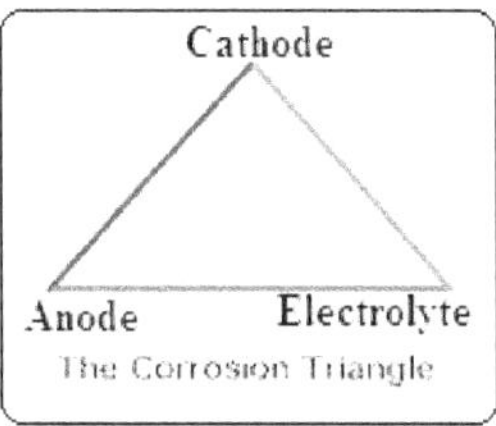

Uniform attack is the most common form of corrosion. It is normally characterized by a chemical or electrochemical reaction, which proceeds uniformly over the entire exposed surface or over a large area.

For example, a piece of steel or zinc immersed in dilute sulfuric acid will normally dissolve at a uniform rate over its entire surface. A sheet iron roof will show essentially the same degree of rusting over its entire outside surface.

Uniform attack can be prevented or reduced by

- proper materials selection including coatings
- inhibitors
- cathodic protection

(ii) Galvanic Corrosion

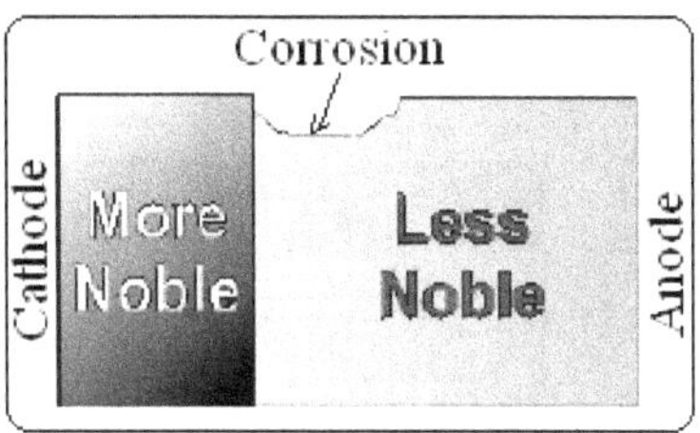

It occurs between two elements having different electrode potential in contact to each other. The metals having higher electrode potential act as anode and the one having less electrode potential acts as cathode. Corrosion occurs in anode.

Galvanic corrosion occurs when two dissimilar metals are connected electrically and are in contact with an electrolyte solution.

One example found in the oilfield is when a new section of pipe is added to an older section. The new pipe becomes anodic and corrodes preferentially.

In general, the further apart the materials are in the galvanic series, the higher the risk of galvanic corrosion.

Electrochemistry of Heat Treatment of Steel

Stainless steel, an alloy of chromium (Cr), nickel and iron, requires at least 12% Cr for passivity.

Stainless steel is heated to a high temperature (such as 425 °C), chromium carbide precipitates will start to form along grain boundaries, leaving a zone depleted of chromium. The precipitates will dissolve back into the grain structure when heated above 850 °C and fast cooled (quenched) back to room temperature.

Stainless steel may become sensitized during welding. The area surrounding the weld bead is known as a heat affected zone (HAZ), a zone depleted of chromium. Therefore, post-welding heat treatment or the use of low-carbon varieties is needed to prevent grain boundary corrosion.

(iii) Crevice Corrosion

Intense localized corrosion occurs within crevices and other shielded areas on metal surfaces exposed to corrosives. This type of attack is usually associated with small volumes of stagnant solution caused by holes, gasket surfaces, lap joints, surface deposits, and crevices under bolt and rivet heads. As a result, this form of corrosion is called crevice corrosion.

(iv) Differential Aeration Corrosion

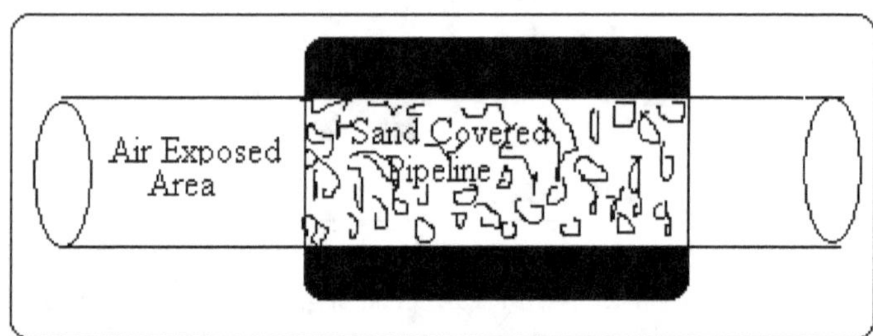

When a metal is exposed in different concentration of air (O_2) differential aeration corrosion occurs. The part, which has limited supply of air acts as anode and gets corroded.

$$Zn \longrightarrow Zn^{2+} + 2e^-$$
$$1/2O_2 + H_2 + e^- \longrightarrow 2OH^-$$
$$Zn^{2+} + 2(OH^-) \longrightarrow n(OH)_2$$

The metals covered with dust, sand, dirt, etc gets corroded faster than clean metal. This is the difference in aeration of the metal surface. This constitutes anodic and cathodic area, which results in corrosion. It is a localized attack of metals.

Accumulation of dirt on metal surface accelerates corrosion due to differential aeration corrosion.

(v) Pitting Corrosion

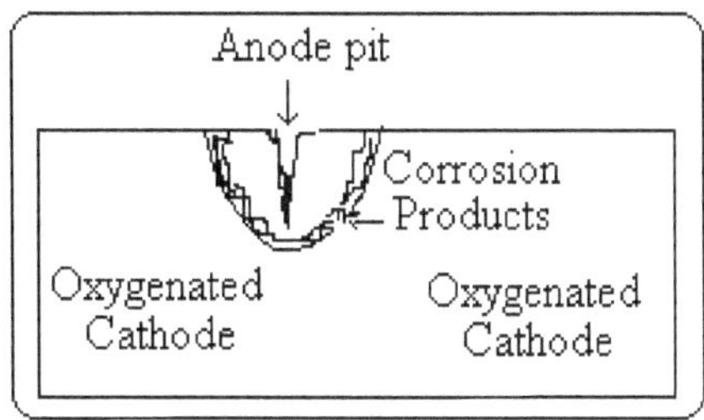

The formation of local cavities, pinholes, pits in localized area of a metal is known as pitting corrosion.

Pitting is a form of extremely localized attack that results in holes in the metal. Pits are sometimes isolated or so close together that they look like a rough surface. Generally a pit may be described as a cavity or hole with the surface diameter about the same as or less than the depth.

Surface roughness, scratches, cuts, local strains, uneven load, etc are the some of the causes of pitting corrosion.

Presence of extraneous impurities like dust, scale or impurity atoms in the metal surface initiates the pitting corrosion. Once it is started the process gets accelerated due to unevenness and goes underneath leading to the formation of pit.

Structural failures, poor performance of tools, decreased efficiency, etc are the consequence of pitting corrosion.

Pitting is one of the most destructive and insidious forms of corrosion. It is often difficult to detect pits because of their small size and because the pits are often covered with corrosion products. In addition, it is difficult to measure quantitatively and compare the extent of pitting because of the varying depths and numbers of pits that may occur under identical conditions.

Pitting is also difficult to predict by laboratory tests. Sometimes the pits require a long time-several months or a year-to show up in actual service. Pitting is particularly vicious because it is a localized and intense form of corrosion, and failures often occur with extreme suddenness

(vi) Intergranular Corrosion

Under certain conditions, grain interfaces are very reactive and intergranular corrosion results. Localized attack at and adjacent to grain boundaries, with relatively little corrosion of the grains, is intergranular corrosion. The alloy disintegrates (grains fall out) and/or loses its strength.

Intergranular corrosion can be caused by impurities at the grain boundaries, enrichment of one of the alloying elements, or depletion of one of these elements in the grain-boundary areas.

Small amounts of iron in aluminum, wherein the solubility of iron is low, have been shown to segregate in the grain boundaries and cause intergranular corrosion. It has been shown that based on surface tension considerations the zinc content of a brass is higher at the grain boundaries. Depletion of chromium in the grain-boundary regions results in intergranular corrosion of stainless steels.

(vii) Selective Leaching

Selective leaching is the removal of one element from a solid alloy by corrosion processes. The most common example is the selective removal of zinc in brass alloys (dezincification). Similar processes occur in other alloy systems in which aluminum, iron, cobalt, chromium, and other elements are removed. Selective leaching is the general term to describe these processes.

(viii) Erosion Corrosion

Erosion corrosion is the acceleration or increase in rate of deterioration or attack on a metal because of relative movement between a corrosive fluid and the metal surface. Generally, this movement is quite rapid, and mechanical wear effects or abrasion are involved. Metal is removed from the surface as dissolved ions, or it forms solid corrosion products which are mechanically swept from the metal surface.

Erosion corrosion is characterized in appearance by grooves, gullies, waves, rounded holes, and valleys and usually exhibits a directional pattern. In many cases, failures because of erosion corrosion occur in a relatively short time, and they are unexpected. This is largely because evaluation corrosion tests were run under static conditions or because the erosion effects were not considered in the tests.

(ix) Stress-Corrosion Cracking

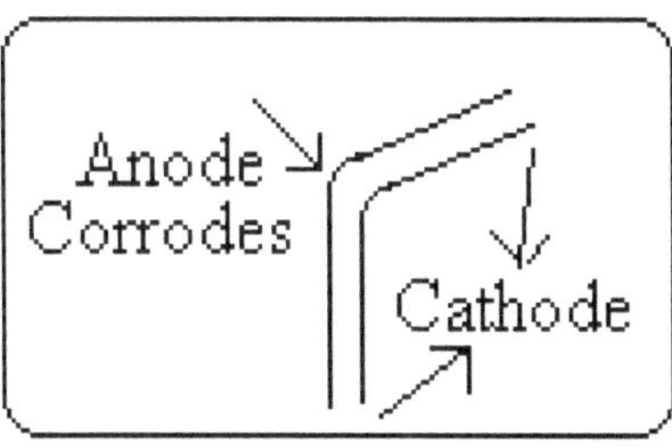

Stress-corrosion cracking refers to cracking caused by the simultaneous presence of tensile stress and a specific corrosive medium. During stress-corrosion cracking, the metal or alloy is virtually unattacked over most of its surface, while fine cracks progress through it. This cracking phenomenon has serious consequences since it can occur at stresses within the range of typical design stress. Exposure to boiling $MgCl_2$ at 27°C (154°C) is shown to reduce the strength capability to approximately that available at 1000°C.

The two classic cases of stress-corrosion cracking are "season cracking" of brass, and the "caustic embrittlement" of steel.

During periods of heavy rainfall, especially in the tropics, cracks were observed in the brass cartridge cases at the point where the case was crimped to the bullet.

Many explosions of riveted boilers occurred in early steam-driven locomotives. Examination of these failures showed the presence of whitish deposits, the showed caustic, or sodium hydroxide, to be the major component. Hence, brittle fracture in the presence of caustic resulted in the term caustic embrittlement.

(x) Microbial Corrosion

Microbial corrosion (also called microbiologically -influenced corrosion or MIC) is corrosion that is caused by the presence and activities of microbes. This corrosion can take many forms and can be controlled by biocides or by conventional corrosion control methods.

Factors Influencing Corrosion

The environmental factors and properties of the metals influence the corrosion potential.

Because of the electric currents and dissimilar metals involved, this form of corrosion is called galvanic, or two-metal corrosion. Mechanistically, it is electrochemical corrosion.

1. Factors Depending on Metal

(i) Purity of Metal

Pure metals are less prone for corrosion. The impurity atoms form tiny electrochemical cells due to the difference in electrode potential of different metal atoms.

For instance zinc containing lead or iron impurity drives the corrosion of zinc. This is the formation of electrochemical cells between Zn and Pb of Fe.

It has been observed that impure zinc (99.95%) is 5000 more prone for corrosion when compared to pure zinc (99.99%).

(ii) Position in Galvanic Series

The corrosion potential depends on the position in the Galvanic series. The metals placed in the beginning are more prone for corrosion. Noble metals are towards the end of the series.

Further if the metals are in contact they undergo Galvanic corrosion. If the metals are farther in the series greater is the corrosion and vice-versa.

(iii) Relative Area of Electrodic Surface

I has been observed that corrosion of anode is directly proportional to the ratio of area of cathodic and anodic parts.

(iv) Solubility of Corrosion Products

Pure metals are least attacked by corrosion due to homogenous array of atoms. Impurity introduces heterogeneity. It leads to localized electrochemical cells and thus corrosion is induced.

(v) Overvoltage

Corrosion and overvoltage are inversely related. In an electrochemical process in which hydrogen is evolved using zinc (active metal) electrode, the overvoltage is 0.70 V. Thus it corrodes slower. However if few drops of $CuSO_4$ solution is added the overvoltage drops to 0.33 V but the corrosion of Zn increases. This is because copper get deposited in zinc surface forming minute cathodes. Thus zinc acts as anode and get corroded.

(vi) Volatility of Corrosion Products

If the corrosion leads to the formation of volatile products the corrosion continues due the exposure of fresh surface for corrosion. For instance molybdenum on oxidation gives volatile molybdenum oxide and hence it continues to undergo corrosion.

In contrast if the corrosion product is covering surface of the metal it decreases the effective surface area available for corrosion. So the corrosion decreases with time.

(vii) Passive Character of the Metal

Passive metals like Tl, Al, C, Mg, Ni and Co shows the tendency of decreasing corrosion on active metals. It is the case where steel is resistant to corrosion due to the presence of passive character of element Cr in it.

(viii) Pilling-Bedworth Rule Consideration

Molecular volume of corrosion products plays a major role in corrosion potential of metals. Porous corrosion products accelerate corrosion. On the other hand if the corrosion product is having higher volume than atomic volume of the element forms a dense surface coating. It prevents further corrosion.

2. Environmental Factors

(a) Humidity

Moisture helps the electrochemical corrosion by providing environment for the formation of electrochemical cells. Further it provides ideal environment for the atmospheric contaminants to dissolve in it and make the corrosion faster by different mechanisms.

(b) Influence of pH

Hydrogen ion concentration has influence in corrosion.

The corrosion of iron in oxygen free water is low until pH = 5. But if oxygen is present the corrosion of iron is high.

Increasing pH can decrease metals, which are more prone to undergo corrosion. Zn gets corroded fast even in weakly acidic solution like carbonic acid. It is suffers least corrosion at pH = 11.

(c) Temperature

Corrosion is also temperature depended process. As the temperature of atmosphere increases there will be increase in corrosion also.

(d) Pollution

Atmospheric pollution enhances the corrosion. Green house gases (SO_2, CO_2, NO_2, etc) increase the corrosion

(e) Suspended Particles

Suspended particles, dusts, etc increase the rate of corrosion. It causes difference in aeration in localized area and thus corrosion occurs.

(f) Differential Exposure of Atmospheric Oxygen

If a metal is exposed to different concentration of atmospheric oxygen it constitute a concentration cell which in turn accelerates the corrosion.

Corrosion Control

1. Better Design

- The contact of two dissimilar metals should be avoided.
- The surface area of more active metal (anode) should be higher than the cathodic metal if there is a contact between two dissimilar metals.
- The metals used must be closely placed in the electrochemical series.
- When a need for connecting two dissimilar metals an insulating fitting can be placed in between them. It avoids the electron flow between the metals.
- Smooth surface as much as possible. Avoid cuts, sharp edges, crevices, etc.
- Stress free design and avoid sharps corners.

2. Use of Pure Metals

Pure metals form a homogenous packing of metal atoms. It prevents the formation of local electrochemical cells and corrosion.

The corrosion rate of 99.95% zinc is 5000 times faster than the pure zinc (99.99%).

3. Use of Alloys

Alloying a metal with other metal can decrease the corrosion provided the alloying element should form a homogenous environment. Chromium is most suitable element for alloying iron.

4. Cathodic Protection

The first application of cathodic protection (CP) can be traced back to 1824, when Sir Humphry Davy, in a project financed by the British Navy, succeeded in protecting copper sheathing against corrosion from seawater by the use of iron anodes.

From a thermodynamics point of view, the application of a CP current basically reduces the corrosion rate of a metallic structure by reducing its corrosion potential towards its immune state.

(i) Sacrificial Anode

The more active metal (anode) is sacrificed to protect the less active metal (cathode). The amount of corrosion depends on the metal being used as an anode but is directly proportional to the amount of current supplied.

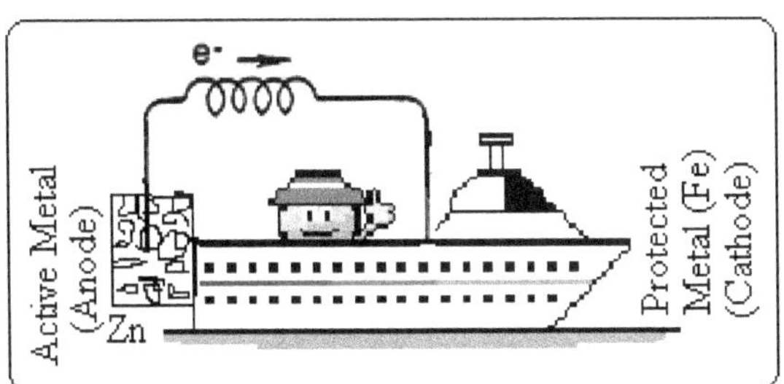

For example, the natural corrosion potential of iron is about -0.550 volts in seawater. The natural corrosion potential of zinc in seawater is about -1.2 volts. Thus if the two metals are electrically connected, the corrosion of the zinc becomes a source of negative charge which prevents corrosion of the iron.

The materials used for sacrificial anodes are either relatively pure active metals, such as zinc or magnesium, or magnesium-aluminum alloys that have been specifically developed for use as sacrificial anodes.

In some applications the anodes are buried, a special backfill material surrounds the anode in order to insure that the anode will produce the desired output.

Merits and Demerits of Sacrificial Anode

Sl.No	Merits	Demerits
1	It is cheaper method to prevent corrosion.	Metal gets destroyed
2	No special device is required.	Environmentally hazardous compounds are formed.
3	No special care is required.	Themselves gets destroyed by the environments since metal used is active one.

(ii) Impressed Current

An external current is applied opposite to the direction of corrosion current. It converts the anode to become cathode and hence prevents corrosion.

Usually DC current is supplied from a battery when want to protect a buried pipe from corrosion.

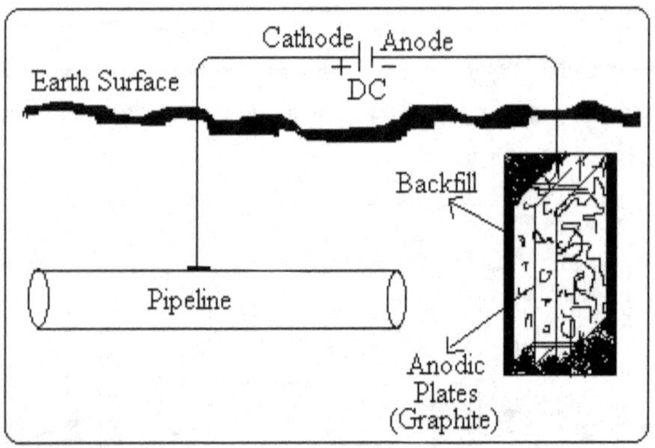

Sl. No.	Merits	Demerits
1	It is a device that can be easily controlled.	It is a costly device.
2	It can be reused many times.	Special care is required.
3	Current supplied can be varied depending upon season, medium, etc.	Corrosion current varies due to many factors. Thus selecting a suitable device and tuning it is required.

5. Corrosion Inhibitors

Corrosion inhibitors are substances added in small quantities in order to prevent or decrease the corrosion of a metal.

(i) Anodic Inhibitors

It cares the corrosion occurring at anode. Its action is generally forming sparingly soluble compound surrounding the metal ion.

Compounds of chromates, tungstates and other transition metal oxides belong to this class of inhibitors.

(ii) Cathodic Inhibitors

In gas evolution processes for example hydrogen evolution occurs at cathode. It involves the migration of H^+ ions.

$$2H^+ + 2e^- \longrightarrow H_2(g)$$

The diffusion of H^+ ions to the cathode can be decreased by increasing hydrogen overvoltage.

Organic compounds like amines, mercaptans, ureas and thioureas are known to have desirable properties in this direction. The mechanism is supposed to be the formation of compounds which is being adsorbed at the metal surfaces. This causes a condition of forming a metallic film at cathodic area thereby decreases the hydrogen overvoltage.

Vapour Phase Inhibitors

In this process compound that can give sufficient vapour is used as inhibitors. The vapour molecules protect the corrosion by forming a protective layer. e.g., benzotriazole, dicyclohexylammonium nitrate.

Protective Coatings

Coating is a surface protection by covering with another material. It isolates the interior from the environment.

The coating must be tolerant to environment. It shall be chemically inert with respect to environment and also to coated surface. It must prevent the penetration of chemicals and to provide protection.

The coating may be inorganic or organic in nature. Metals and non-metals are coated by special techniques like electroplating, electoless coating, ion implantation, thermal spraying, etc.

Paints, varnishes, etc are organic coatings.

(i) Anodic Coating

The coated material acts as anode in relation to the base metal on which the coating is applied. It indicates the coating material has lower electrode potential than the base metal.

Mechanistically the base metal acts as cathode and being protected. The coating acts as anode and suffers corrosion.

The distinction from sacrificial anode is coating is very uniform minute layer of few micron thickness covering entire surface area of the base metal.

A coating of metals like Zn, Al, Cd, etc over steel is an example of anodic coating. The coating should be as dense as possible and uniform thereby prevents the interior metal being exposed to corrosion devil.

(ii) Cathodic Coating

It is the process of coating a surface area to be protected by a noble metal like platinum, gold, etc. Since these metals are having higher electrode potential and has higher corrosion resistance against the factors which causes corrosion. The effective protection is possible only when a complete coating on entire surface of the base metal. Thus it is a costly process of using precious metals. If any defect in coating it sets up a Galvanic cell and the corrosion is accelerated many fold.

Difference between Cathodic and Anodic Protection

Sl.No.	Cathodic Protection	Anodic Protection
1	Coating material has high electrode potential than base metal.	Coating material has lower electrode potential than base metal.
2	Protection is due to high degree of corrosion resistance of the coating.	Protection is due to sacrificial corrosion of the coating.
3	The process of corrosion is prevented.	The process does exist but occurs on the coating.
4	e. g., Sn coating on Fe surface	e. g., Zn coating on Fe surface.

Paints

Paints are organic compounds generally pigments which are used to coat on the metallic surfaces. It provides corrosion protection in addition to aesthetic and decorative values. Paint is a colloidal suspension in some liquid. It consists of volatile liquid and non-volatile pigment. A good paint must posses the following characteristics.

- It is being capable of forming thin film.
- Must have desired corrosion protection.
- Should not react with the metals causing undesirable effect.
- It must have adequate adhering property.
- The paint film must be stable for long time.
- It must support the coating techniques and process.
- Sufficiently miscible with a suitable solvent.

Constituents of Paints

Hence paint is a homogeneous mixture of many components it has many constituents. Few are discussed below.

(a) Pigment

It is the colouring part of the constituent and is the very important one in a paint. The colour provided by the paint is coming from pigment. It is generally the major constituent other than solvent. Pigment is in its isolation is a solid compound. A good pigment must give good appeal to paint. It is required to give protection to corrosion. Longer the life of the pigment better is the quality of paint. It should not fade away by harsh environment. It should give in lowest possible concentration a brightest appearance.

Some of the traditional pigments include

Inorganic pigments like red-lead, ferric oxide, chrome red, etc and organic ones like prusian blue, carbon black, etc.

Pigments may be white in colour (white lead, zinc oxide, etc) or coloured (black carbon, chromium oxide, red-red lead, ferric oxide, green-chromium oxide).

(b) Vehicle

It is the medium in which paint is dispersed. It constitutes one of the major parts of the paint and one involved in thin film formation on coating. It has to accommodate the constituents and keep them in the liquid state for the desired period of time. In general they are organic compounds.

The vehicle has to give film within sufficient period time with out leaving pores or wholes on the paint surface.

Fatty acid esters, caster oil, aliphatic and aromatic hydrocarbons are used as vehicle in many paints. It also contains reactive functional groups like double bonds.

Vehicle has to be reactive, could initiate or undergo chemical reactions. Generally they participate in radical polymerization using atmospheric oxygen with the aid or without the aid of initiators forming three dimensional crosslinks in which pigments are embedded into it.

(c) Thinners

Thinners are solvents. Its desirability in paints is due its function of reducing viscosity of the paint as a whole thereby allows to apply it in relatively easy way on the surface. It enhances the spreading area and thus an economic contribution too is involved.

It has its own suitability in allowing pigments, vehicle and other constituents in its desired liquid state sufficiently enough for application on surfaces. It enhances the quality of vehicle in terms of uniform spreading, thin continuous film formation etc. It ensures easy drying and provides required elasticity. It constitutes in general organic solvents which includes turpentine, petroleum, benzene, kerosene, etc. Recent times water soluble paints and pigments are in the market.

(d) Driers

There are oxygen container catalysts. It supplies sufficient assistance for polymerization process after painting on a surface.

Drier can be organic or organic compounds. Few examples include linolates, resins, cobolt and lead compounds.

(e) Fillers

These are chief materials added to reduce the cost but increase the volume of the paint. They either need or need not interact with the other constitutes of the paint but should involve in giving undesirable quality to paint.

It provides supports in filling voids developed while drying the paints. It may help the pigments in suspended state. It also provides durability for paints.

Talc, asbestos, gypsum, clay calcium suphate , etc. are some of the fillers widely used in the paint industry.

(f) Platicizers

The materials added in the paint to enhance the plasticity is called plasticizers.

The examples include triphenyl phosphate, tricreyl phosphate, diamyl phosphate, etc.

Functions of Paints

- Protects from corrosion.
- It gives aesthetic value to the article.
- It improves quality of the articles.
- It enhances economic value of the materials.
- It serves as decorative purpose.
- It increases the life time of the instruments.
- It prevents environmental contamination.

Metallic Coatings

Metallic coatings provide a layer that changes the surface properties of the work piece.

The coatings provide a durable, corrosion resistant layer, and the core material provides the load bearing capability. The deposition of metal coatings, such as chromium, nickel, copper, and cadmium, is usually achieved by wet chemical processes.

Metallic coatings are deposited by electroplating, electroless plating, spraying, hot dipping, chemical vapor deposition ion implantation and ion vapor deposition. Some important coatings are cadmium, chromium, nickel, aluminum and zinc.

Electroplating (Au)

Italian chemist, Luigi Brugnatelli invented electroplating in 1805. Luigi Brugnatelli's work was rebuffed by the dictator Napoleon Bonaparte, which caused Brugnatelli to suppress any further publication of his work.

Forty years later, John Wright of Birmingham, England discovered that potassium cyanide was a suitable electrolyte for gold and silver electroplating.

Principle of Electroplating

Electroplating can be defined as the deposit of a very thin layer of metal "electrolytically" to a base metal.

Electroplating is done in a liquid solution called an electrolyte, also known as a "plating bath". The desired metal (i.e. silver, gold) dissolved as microscopic particles (positive charged ions) suspended in solution. The plating bath solution serves as a conductive medium and utilizes a low d.c. voltage (direct current). The object that is to be plated as cathode and is submerged into the plating bath.

The electrical current flows from positive to negative terminal. The positively charged metal ions in the electrolyte move toward the negatively charged cathode and are deposited as a thin layer onto the surface of the object. This process is called electrodeposition.

Ferrous and non-ferrous metal objects are plated with a variety of metals, including aluminum, brass, bronze, cadmium, copper, chromium, iron, lead, nickel, tin, and zinc, as gold, platinum, and silver. Elements such as Ti and Al can only be deposited from organic electrolytes, while other metals such as Mg, Nb, Ta, and W can only be plated from molten salt electrolytes (at 700°C and above).

Factors Influence the Electroplating

- Concentration of electrolyte.
- Current density.
- Temperature of the plating bath.
- pH of the medium.

Functions of Electroplating

- Corrosion protection
- Improvement of aesthetic qualities
- Shining
- Improves the abrasion resistance and wear.

Electroplating of Gold

Gold is unique with its yellow color. Also, gold is a precious metal, which means that it will not oxidize in air, so its electrical conductivity stays uniform over long periods of time. It is ideally suited for gold electroplating applications. Gold plating offers good corrosion resistance, good solderability, and when alloyed with cobalt, it has very good wear resistance.

Gold is commonly used in electrical switch contacts, connector pins and barrels, and other applications where intermittent electrical contact occurs.

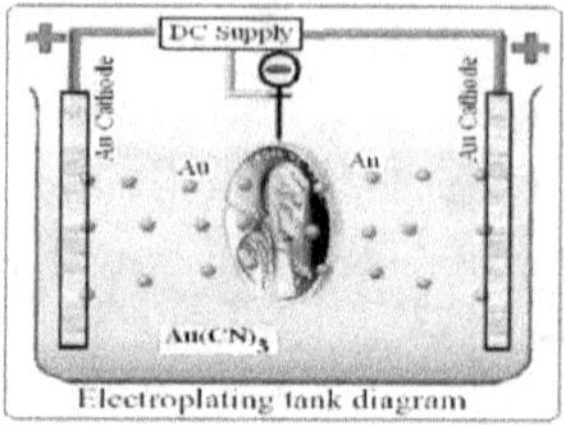

Electroplating tank diagram

Components of Gold Electroplating

	Component	Quantity	Function
1	Gold as Gold cyanide.	4-12 g/l	Coating metal
2	Citric Acid	20-70 g/l	Buffer
3	Potassium Citrate	50-90 g/l	Buffer
4	Cobalt as sulphate	1-3 g/l	Wear and tear resistance.

Operating temperature between 29-35 °C.

pH is maintained between 3.2-4.0.

Current density is 0.8-2 A/dm².

Cobalt is used along with gold in order to reduce the cost as prime reason in addition to alloying element. It provides wear and tear resistance.

Mechanism of Gold Plating

- Electrolysis of Gold Cynanide

$$Au(CN)_3 \rightarrow Au^{3+} + 3CN^-$$

- Reduction at Anode

$$Au^{3+} + 3e^- \rightarrow Au$$

  The gold formed is deposited on the surface of the metal cathode.

- Oxidation at Cathode

$$Au^{3+} + 3CN^- \rightarrow Au(CN)_3$$

  Thereby gold anode gets dissolved into solution

Functions of Gold Platting

- Gold is a noble metal. So it is resistant to corrosion.
- It is a good conductor of heat and electricity and thus it improves the thermal and electrical property of the base metal.

- In gives value addition to the base metal.
- It give a aesthetic satisfaction to people.

Electroless Plating (Ni)

The term electroless plating was originally adopted by Brenner and Riddell to describe a method of plating metallic substrates with nickel or cobalt alloys without the benefit of an external source of electric current.

In general, electroless plating is characterized by the selective reduction of metal ions only at the surface of a catalytic substrate immersed into an aqueous solution of said metal ions, with continued deposition on the substrate through the catalytic action of the deposit itself.

Principle

It is an electroplating technique in which noble metal is coated on a less noble metal surface without the use of electric current. In this method coating is achieved by use reducing agent. It causes reduction of noble metal ions from its solution and in the process base metal is coated.

$$\text{Metal ions} + \text{Reducing Agent} \longrightarrow \text{Metal} + \text{Oxidized Product}$$

Electroless nickel (EN) plating is a chemical reduction process which depends upon the catalytic reduction process of nickel ions in an aqueous solution (containing a chemical reducing agent) and the subsequent deposition of nickel metal without the use of electrical energy. Due to its exceptional corrosion resistance and high hardness, the process finds wide application on items such as valves, pump parts etc., to enhance the life of components exposed to severe conditions of service, particularly in the oil field and marine sector.

Components of Electroless Nickel Plating

Source of Nickel	Nickel Sulphate, $NiCl_2$	20 g/l
Reducing Agent	Sodium hypophosphite	20 g/l
Complexing Agent	Sodium succinate	100-120 mg/l
Stabilizer	Maleic acid, MoO_4^{2-}, etc.	20-25 mg/l
Buffer	Sodium citrate	0.1 M to 0.5 M

The process is carried at a pH of 4.5 and temperature 93-95 °C.

Mechanism of Electroless Nickel Coating

It occurs due to redox process as follows.

$$Ni^{2+} + 2e^- \longrightarrow Ni$$

$$H_2PO_2^- + H_2O \longrightarrow H_2PO_3^- + 2H^+ + 2e^-$$

$$Ni^{2+} + H_2PO_2^- + H_2O \longrightarrow Ni + H_2PO_3^- + 2H^+ \text{ (Net reaction)}$$

In the EN plating process, the driving force for the reduction of nickel metal ions and their deposition is supplied by a chemical reducing agent in solution.

Functions of Electroless Nickel Plating

Electroless nickel (EN) plating is known and is useful for generating corrosion and wear resistant coatings on metal or plastic parts, both on open surfaces or inside cavities.

It protects electronic appliance from corrosion. EN is employed in computers, valves, aircraft parts, copier and typewriter parts and printing press rolls.

Electroless deposits offers very uniform in thickness all over the parts irrespective of shape and size.

This process offers distinct advantages when plating irregularly shaped objects, holes, recesses, internal surfaces, valves or threaded parts.

Advantages of Electroless Plating

- Uniformity of the deposits, even on complex shapes.
- Deposits can be plated with zero compressive stress.
- Deposits have good wetability for oils
- Deposits are often less porous and thus provide better barrier corrosion protection to steel substrates, much superior to that of electroplated nickel and hard chrome.
- Deposits have inherent lubricity and non-galling characteristics, unlike electrolytic nickel.
- Improves solderability characteristics.
- Deposits are much harder and thus suitable heat-treatment

Comparison of Electroplating and Electroless Plating

Sl. No.	Electroplating	Electroless Plating
1	It is an electrolysis process by an external current.	No external current supply. Only internal redox reaction.
2	Quality of coating is poor and not uniform.	Very good quality and uniform.
3	Less suitable for complex irregular shape articles	Any complex article can be plated successfully.
4	Anode and cathode are electrodes.	Catalyst surface acts as anode.
5	The plating surface must be electrically conducting.	Can be employed for semiconducting materials.

Metal Cladding

Cladding is by pressing, rolling or extrusion of a metal sheet on a surface of an another sheet.

Indeed, there is almost no practical limit to the thickness of coatings that can be produced by cladding, but decided depending upon applications.

Cladding by anodic metal will provide cathodic protection by corroding sacrificially.

Cladding can applied at the mill stages by the manufacturers of sheet, plate or tubing.

Physical Vapour Deposition (PVD)

Physical vapor deposition methods are clean, dry vacuum deposition methods in which vapour of a metal is deposited over the entire object.

The primary PVD methods are ion plating, ion implantation, sputtering, and laser surface alloying.

Assignment Questions

1. Make a comparative analysis of electochemical series and Galvanic series.
2. Differentiate electroplating and electroless plating.
3. Distinguish impressed current and sacrificial anode type protection of corrosion.
4. Explain the role of paints in prevention of corrosion.
5. Distinguish corrosion current and electrode potential.

Self Test

1. Define corrosion and give its impact on human life.
2. Bring out the salient features of different types of corrosion.
3. What are physical factors influence the corrosion? Explain.
4. Explore the characteristics required for a metal to act as an sacrificial anode.
5. Bring a correlation between corrosion current and impressed current.
6. Mechanistically explain the role of corrosion inhibitors in the prevention of corrosion.
7. Define coating and coating process. Differentiate coating from painting.
8. What are characteristics of which are relevant in protection and prevention of corrosion.
9. Explain the principle and mechanism of electroless plating.
10. Explore and explain briefly few non-electrolytic coating processes.

UNIT III

FUELS AND COMBUSTION

Calorific value-classification-coal-proximate and ultimate analysis, metallurgical coke-manufacture by Otto-Hoffmann method-petroleum processing and fractions, cracking-catalytic cracking and methods knocking-octane number and cetane number- synthetic petrol- Fischer-Tropsch and Bergius processes. Gaseous fuels- water gas, producer gas, CNG and LPG, Flue gas analysis, Orsat apparatus, theoretical air for combustion.

Fuel

Fuel is defined as a combustible substance containing major constituent as carbon. On combustion it produces large amount of heat, which is used for various process where heat energy required. eg. Coal, petroleum, etc.

Combustion is an exothermic process in which fuel burns in oxygen. The major product of combustion of a fuel are CO_2 and H_2O.

$$\underset{12}{C} + O_2 \longrightarrow \underset{44}{CO_2}\uparrow$$

$$\underset{2}{H_2} + 0.5\,O_2 \longrightarrow \underset{18}{H_2O}$$

Calorific Value

The quantity of heat liberated when unit mass or volume of a fuel is burnt completely in the presence of air or oxygen. Higher the calorific value better the quality of the fuel. Calorific value is determined in bomb calorimeter.

Units of Calorific Value

(i) Calorie

SI unit of calorific value is calorie. It is the quantity of heat required to raise the temperature of one kilogram of water by $1\,^{\circ}C$.

(ii) British Thermal Unit (BTU)

It is the quantity of heat required to raise the temperature of one pound of water by $1\,^{\circ}F$. It is generally used for solid and liquid fuels.

(iii) Centigrade Heat Unit (CHU)

It is the "quantity of heat required to raise the temperature of one pound of water by $1\,^{\circ}C$."

Relation among Units of Calorific Value

1 kcal = 3.968 BTU = 2.2 CHU

1 kcal/kg = 1.8 x BTU/lb.

1 kcal/m³ = 0.1077 x BTU/ft³

1 BTU/ft³ = 9.3 kcal/m³.

Higher or Gross Calorific Value (HCV or GCV)

It is the calorific value measured when unit mass or volume of fuel is burnt completely and the products of combustion are cooled to room temperature (1ºC). In HCV latent heat of condensation is also included in the measured value.

Lower or Net Calorific Value (LCV or NCV)

It is the calorific value measured when unit mass or volume of fuel is burnt completely and the products of combustion are allowed to escape. Thus LCV is always lower than HCV.

LCV = HCV – Latent heat of vaporization of water.

Theoretical Calculation of Calorific Value

According to Dulong's formula theoretical calorific value can be calculated using the following formula.

$$HCV = \frac{1}{100}\left[8080 \text{ x C} + 34500 \left(H - \frac{O}{8}\right) + 2240 \text{ S}\right] \text{ kcal/kg}$$

C, H and S stands for the percentage of respective elements in the fuel. It is based on the assumption that calorific value of C, H and S is 8080, 34500 and 2240 kcal respectively.

LCV = (HCV – 0.09 x % H x 587) kcal/kg

587 is the latent heat of vaporization of water.

Problem

1. Calculate gross and net calorific value of coal having the following composition. C = 85 %, H = 8 %, S = 1 %, and ash = 4%.

   Gross Calorific value

   $$HCV = \frac{1}{100}\left[8080 \text{ x C} + 34500 \left(H - \frac{O}{8}\right) + 2240 \text{ S}\right] \text{ kcal/kg}$$

   $$HCV = \frac{1}{100}\left[686,800 + 276,000 + 2240\right] \text{ kcal/kg}$$

$$HCV = \frac{1}{100}[686,800 + 276,000 + 2240] \text{ kcal/kg} = 9,650.4 \text{ kcal/kg}$$

Net calorific value = GCV – (0.09H x 587) kcal/kg

$$= 9.650.4 - (0.09 \times 8 \times 587) = 9,227.8 \text{ kcal/kg}.$$

2. A coal has the following composition by weight. C = 90 %, O = 3 %, S = 0.5 %, N = 0.5 % and ash = 2.5 %. Net calorific value is 8,490.5 kcal/kg. Calculate the percentage of hydrogen and high calorific value.

LCV = (HCV - 0.09 x H x 587) kcal/kg

HCV = (LCV + 0.09 x H x 587) kcal/kg

H CV = (8,490.5 + 52.8 x H) kcal/kg ----------- (i)

$$HCV = \frac{1}{100}\left[8080 \times 90 + 34500\left(H - \frac{3}{8}\right) + 2240 \times 0.5\right] \text{ kcal/kg}$$

$$= 7,754.8 \times 345 \text{ H} \text{ ------- (ii)}$$

On solving equation (i) and (ii)

7,754.8 x 345 H = 8,490.5 + 52.8 x H

292.2 H = 8,490.5 – 7,154.8 = 1,335.7

$$\text{Percentage of H} = \frac{1,335}{292.2} = 4.575 \text{ \%}$$

HCV = 8,490.5 + (52.8 x 4.575) = 8,731.8 kcal/kg

Characteristic of a Good Fuel

- High calorific value.
- Moderate ignition temperature.
- Low ash content.
- Low moisture content.
- Least emission of environmentally harmful gases.
- Low cost.
- Easy to transport and storage.

Classification of Fuels

1. Based on occurrence fuels are classified as natural or artificial fuels.

(i) Natural or Primary Fuels

These are available in nature on surface and inside earth. eg. Wood, coal, petroleum, natural gas, etc.

(ii) Artificial or Secondary Fuels

These are man made fuels by synthetic methods. eg. Charcoal, coke, diesel oil, producer gas, gober gas, etc.

2. Based on the physical state fuels are classified as solid, liquid and gaseous fuels.

(a) Solid Fuels

Solid in nature at room temperature and pressure. eg. Wood, coal, coke, charcoal, etc.

(b) Liquid Fuels

Liquid in nature at room temperature and pressure. e.g., Petroleum, gasoline, diesel oil, etc.

(c) Gaseous Fuels

Gaseous in nature at room temperature and pressure. eg. Natural gas, water gas, biogas, etc.

Sl. No	Origin	Solid	Liquid	Gas
1	Natural	Wood, Lignite	Petroleum Oil	Natural Gas
2	Synthetic	Charcoal, Coke	Biodiesel, Gasoline	Coal Gas, Biogas

Solid Fuels-Coal

Coal is a highly carbonaceous matter. It is a primary fuel and thus naturally occurring. It is formed as a result of vegetable matter under extreme natural conditions of temperature and pressure. It is composed of C, H, N and O with some non-combustible inorganic matter.

Types of Coal

Coal is classified based on their carbon content.

Sl. No.	Name	Carbon Content	Calorific Value Kcal/kg
1	Peat	57 %	5400
2	Lignite	67 %	7000
3	Sub-bituminous	77 %	7000
4	Bituminous	83 %	8300
5	Semi-bituminous	90 %	8500
6	Anthracite	93 %	8700

Analysis of Coal

A. Proximate Analysis

It is a preliminary analysis of coal.

(i) Moisture Content

It is a measure of water content of the coal. It is weight loss when 1g finely powdered coal is heated to 105-110ºC for 1 hour in an oven.

$$\% \text{ Moisture content } = \frac{\text{Loss in weight}}{\text{Weight of coal taken}} \times 100$$

(ii) Volatile Matter

It is a measure of volatile components in a coal. It is the weight loss when 1 g of dried coal is heated to 950ºC for 7 minutes in a muffle furnace with a covered lid.

$$\% \text{ Volatile matter} = \frac{\text{Loss in weight}}{\text{Weight of coal taken}} \times 100$$

(iii) Ash Content

It is a measure of residual content in a coal. It is the weight loss when 1 g of dried coal is heated to 700ºC in a muffle furnace with an open container till constant weight is obtained.

$$\% \text{ Ash content} = \frac{\text{Weight of ash}}{\text{Weight of coal taken}} \times 100$$

(iv) Fixed Carbon

It is determined by subtracting moisture content, volatile matter and ash content from 100.

% Fixed carbon = 100 – (moisture content + volatile matter + ash content)

Significance of Proximate Analysis

Higher the moisture content lower the calorific value and vice versa. However upto 10% moisture is desirable since it reduces 'fly-ash'.

Higher volatile matter indicates that large portion escapes as unburnt and thus undesirable. It burns with a long smoky flame and have low calorific value. Coal containing 26-30% volatile matter is suitable for coke manufacture since it cakes well.

Higher the ash content lowers the quality. It forms clinkers and causes trouble like blocks air supply and prevents regular burning of the fuel.

Higher the percentage of fixed carbon better the quality and higher the calorific value. It helps in designing furnace since it burns in solid state.

1. Ultimate Analysis

It is the quantitative determination of carbon, hydrogen, nitrogen, oxygen and sulphur.

(a) Carbon and Hydrogen Analysis

Accurately weighed (0.2 g) sample is burnt in a current of oxygen in a combustion apparatus. The gaseous products are absorbed in KOH and $CaCl_2$ tubes of known weight.

$$\% \text{ of Carbon} = \frac{\text{Increase in weight of KOH tube x 12}}{\text{Weight of coal taken x 44}} \times 100$$

$$\% \text{ of Hydrogen} = \frac{\text{Increase in weight of } CaCl_2 \text{ tube x 2}}{\text{Weight of coal taken x 18}} \times 100$$

$$C + O_2 \longrightarrow CO_2 \uparrow$$
$$12 \qquad\qquad\qquad 44$$
$$H_2 + 0.5O_2 \longrightarrow H_2O$$
$$2 \qquad\qquad\qquad 18$$
$$2KOH + CO_2 \longrightarrow K_2CO_3 + H_2O$$
$$CaCl_2 + 7H_2O \longrightarrow CaCl_2\,7H_2O$$

(b) Nitrogen Analysis

Accurately weighted (1 g) powdered coal is heated with conc. H_2SO_4 with K_2SO_4 catalyst in a Kjeidahl flask till the solution becomes clear. It is further treated with excess KOH and the liberated ammonia is distilled over and absorbed in a known volume of standard acid. The unused acid is determined by back titration using standard NaOH.

$$\% \text{ of Nitrogen} = \frac{\text{Volume of acid used x Normality x 14}}{\text{Weight of coal taken}} \times 100$$

$$N_2 \text{ (in coal)} + H_2SO_4 \xrightarrow{K_2SO_4} (NH_4)_2SO_4$$
$$(NH_4)_2SO_4 + 2KOH \longrightarrow 2NH_3 + K_2SO_4 + 2H_2O$$
$$NH_3 + HCl \longrightarrow NH_4Cl$$

(c) Sulphur

Known amount of coal (1 g) is burnt in a bomb calorimeter. The sulphates formed are washed with $BaCl_2$ solution. The precipitate $BaSO_4$ (weight 233) formed is filtered, washed and dried to constant weight.

$$\% \text{ of Sulphur} = \frac{\text{Weight of } BaSO_4 \text{ x 32}}{\text{Weight of coal taken x 233}} \times 100$$

$$Fe_2S_3 + 4O_2 \longrightarrow 2FeO + 3SO_2$$
$$4FeO + O_2 \longrightarrow 2Fe_2O_3$$

(d) Ash Analysis

It is carried out as in proximate analysis.

(e) Oxygen Analysis

% of Oxygen = 100 – Percentage of (C + H + S + N + ash)

Significance of Ultimate Analysis

Classification of coal is based on carbon content. Greater is the carbon content smaller is the combustion chamber required.

Higher the carbon and hydrogen content better the quality and greater the calorific value.

Presence of nitrogen is undesirable in coal. It has no calorific value.

Sulphur occurs in both organic and inorganic form. Presence of sulphur contaminates the coke and badly affects its use in metallurgy. Further the oxides of sulphur in combustion process have great environmental effect.

High oxygen content coal has greater moisture content and low calorific value. Thus a good quality coal should have lower percentage of oxygen.

Carbonization of Coal

When coal is heated strongly in the absence of air (destructive distillation), it is converted into lustrous, dense, porous and coherent mass called coke. The process of converting coal to coke is called carbonization.

Low temperature carbonization is carried out at 500-700°C. The coke formed is dense and cannot be used for metallurgy.

High temperature carbonization is carried out at 900-1200°C. The product has high volume and used for metallurgy.

Sl. No	Low Temperature Carbonization	High Temperature Carbonization
1	500-700°C	900-1200°C
2	Mechanically not strong	Good Mechanical strength
3	Used for domestic purpose.	Used in Metallurgy
4	130-150 m³/tonne by-product formed	300-390 m³/tonne by-product formed
5	Calorific value 6500-9500 cal/m³	Calorific value 5400-6000 kcal/m³

Metallurgical Coke

Coke is an artificial solid fuel manufactured by destructive distillation of bituminous coal. It is used as a reducing agent in metallurgy like manufacture of steel.

For the purpose of metallurgy, the coke should be highly pure, porous, free from impurities like sulphur and phosphorous with low reactivity having high combustion temperature.

Metallurgical coke can be manufactured either by Beehive oven process or by Otta-Hoffman process.

Manufacture of Metallurgical Coke (Otta-Hoffman Process)

It is a by product recovery process in addition to the production of coke. In silica chambers coal is charged and heated to 1200ºC using preheated air and producer gas mixture.

The hot flue gases produced during carbonization circulated through regenerators and maintain the temperature at about 1000ºC. Further for economy of heating the inlet gases and flue gases are recirculated and also the direction of flow changed.

Carbonization takes about 15 hours and the yield is about 70%. The flue gases contain ammonia, naphthalene, benzene, hydrogen suphide, tar, etc. They are recovered independently and used for many purposes.

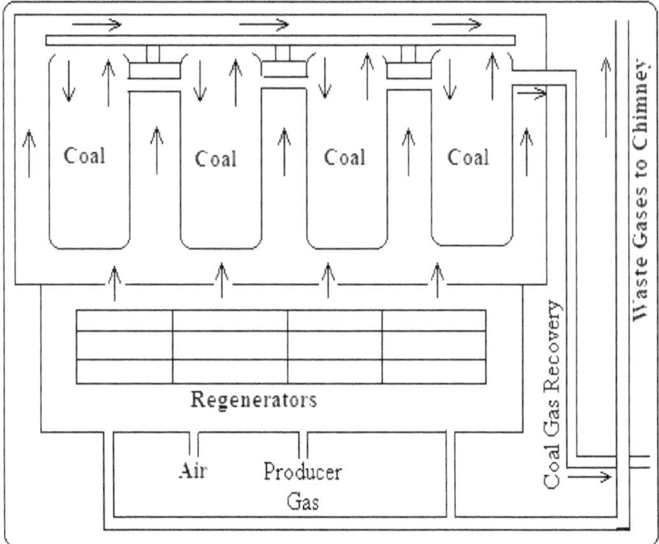

Recovery of By-Products

Tar is recovered using liquor ammonia. The ammonia is distilled out and reused. The tar remains as residue. Ammonia gas is recovered as ammonium hydroxide by passing through water. Recovery of naphthalene is passing through a tower where cooled water is sprayed. It gets condensed as crystals.

Benzene is recovered by spraying petroleum where it dissolved and collected as a liquid.

Hydrogen sulfide is passed through a purifier packed with Fe_2O_3 where H_2S is absorbed as Fe_2S_3.

Then the purifier is exposed to atmosphere where Fe_2O_3 is regenerated.

$$Fe_2O_3 + 3H_2S \longrightarrow Fe_2S_3 + H_2O$$

$$Fe_2S_3 + 4O_2 \longrightarrow 2FeO + 3SO_2$$
$$4FeO + O_2 \longrightarrow 2Fe_2O_3$$

Advantages and Disadvantages of Solid Fuels

Sl.No.	Advantages	Disadvantages
1	Convenient to store	Large amount of ash
2	Easy to transport	Occupies more space
3	Controlled combustion	Burn with clinker formation
4	Moderate ignition temperature	Not suitable for internal combustion engine
5	Cheap and available.	Low calorific value.

Liquid Fuels-Petroleum

Petroleum is a complex mixture of organic liquids called crude oil and natural gas. It occurs naturally in the ground called crude oil. The composition and colour of crude oil varies from oilfield to oilfield.

Structures of Petroleum Hydrocarbon

Petroleum consists of three main hydrocarbon groups:

Paraffins

These consist of straight or branched carbon rings saturated with hydrogen atoms, the simplest of which is methane (CH_4) the main ingredient of natural gas. Others in this group include ethane (C_2H_6), and propane (C_3H_8).

Naphthenes

Naphthenes consist of carbon rings, sometimes with side chains, saturated with hydrogen atoms. Naphthenes are chemically stable, they occur naturally in crude oil and have properties similar to paraffins.

Aromatics

Aromatic hydrocarbons are compounds that contain a ring of six carbon atoms with alternating double and single bonds and six attached hydrogen atoms. This type of structure is known as a benzene ring.

Atmospheric distillation takes place in a distilling column at or near atmospheric pressure. The crude oil is heated to 350-400°C. The vapour and liquid are piped into the distilling column. The liquid falls to the bottom and the vapour rises, passing through a series of perforated trays (sieve trays). Heavier hydrocarbons condense more quickly and settle on lower trays and lighter hydrocarbons remain as a vapour longer and condense on higher trays.

Petroleum Processing and Fractions

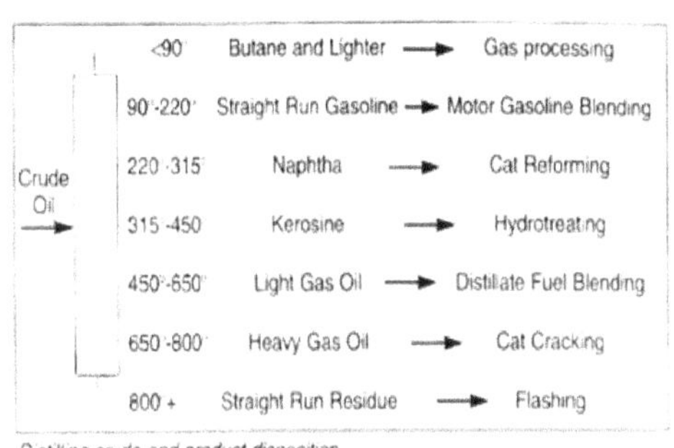

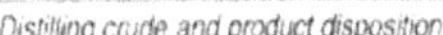
Distilling crude and product disposition

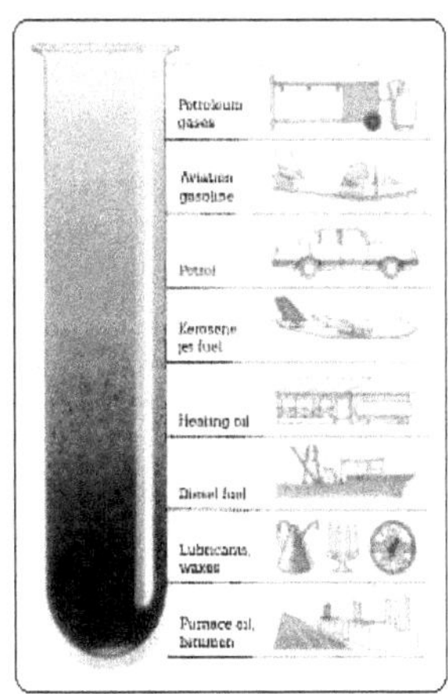

Vacuum distillation allows the recovery of heavy hydrocarbons with boiling points of 450°C and higher hydrocarbons. Heavy oil fraction is subjected to catalytic cracking to convert into gasoline.

The heavy distillates recovered by vacuum distillation can be converted into lubricating oils by a variety of processes.

Cracking

"Cracking" breaks larger molecules into smaller ones. This can be done with a thermic or catalytic method.

Cracking processes break down heavier hydrocarbon molecules (high boiling point oils) into lighter products such as petrol and diesel.

Methods of Cracking

1. Thermal Cracking

Low Temperature Cracking

William Merriam Burton in 1912 developed the earliest thermal cracking processes. It operated at 700-750 °F (370-400 °C) and an absolute pressure of 620 kPa. It known as the Burton process.

High Temperature Cracking

In 1921, C.P. Dubbs developed a somewhat more advanced thermal cracking process which operated at 750-860°F (400-460°C) and was known as the Dubbs process.

The thermal cracking process follows a homolytic mechanism, that is, bonds break symmetrically and thus pairs of free radicals are formed.

$$H_3C\!-\!CH_3 \longrightarrow 2CH_3^{\bullet}$$

In steam cracking, initiation usually involves breaking a chemical bond between two carbon atoms.

$$2CH_3^{\bullet} + H_3C\!-\!CH_2^{\bullet} \longrightarrow H_3C\!-\!CH_2\!-\!CH_3$$

Hydrogen abstraction, where a free radical removes a hydrogen atom from another molecule, turning the second molecule into a free radical.

$$H_3C\!-\!CH_2^{\bullet} \longrightarrow H_2C\!=\!CH_2 + H^{\bullet}$$

Radical decomposition, where a free radical breaks apart into two molecules, one an alkene, the other a free radical. This is the process that results in the alkene products of steam cracking.

Radical addition, the reverse of radical decomposition, in which a radical reacts with an alkene to form a single, larger free radical. These processes are involved in forming the aromatic products that result when heavier feedstocks are used.

Termination reactions, which happen when two free radicals react with each other to produce products. Two common forms of termination are recombination, where the two radicals combine to form one larger molecule, and disproportionation, where one radical transfers a hydrogen atom to the other, giving an alkene and an alkane.

Catalytic Cracking

The catalytic cracking process involves the presence of acid catalysts (usually solid acids such as silica-alumina and zeolites). It promotes a heterolytic (asymmetric) breakage of bonds yielding pairs of ions. They are highly unstable and undergo processes of chain rearrangement, C-C scission in position beta (i.e., cracking) and intra- and intermolecular hydrogen transfer or hydride transfer.

(a) Fluid Bed Catalytic Cracking

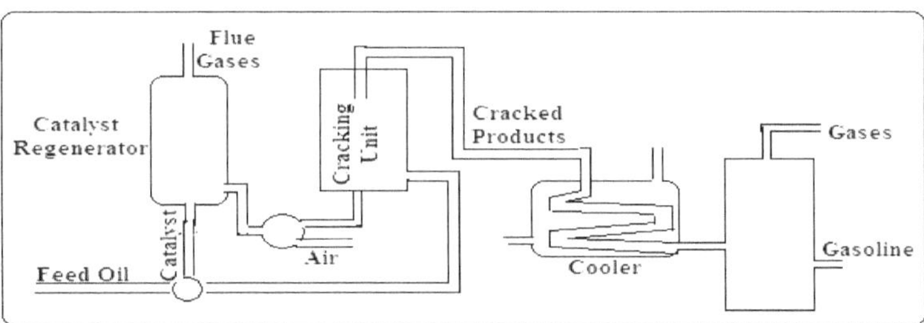

It uses a catalyst in the form of a very fine powder which flows like a liquid when agitated by steam, air or vapour. Feedstock entering the process immediately meets a stream of very hot catalyst and vaporises. The resulting vapours keep the catalyst fluidised as it passes into the reactor, where the cracking takes place.

The catalyst next passes to a steam stripping section where most of the volatile hydrocarbons are removed. It then passes to a regenerator vessel where it is fluidized by a mixture of air and the products of combustion which are produced as the coke on the catalyst is burnt off. The catalyst then flows back to the reactor.

The catalyst is usually a mixture of aluminum oxide and silica. Recent time synthetic zeolite catalysts have allowed much shorter reaction times and improved yields and octane numbers of the cracked gasoline.

(b) Fixed Bed Catalytic Cracking

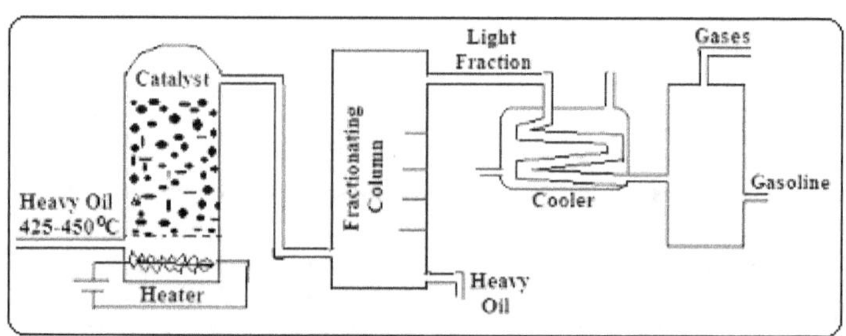

The catalyst silica alumina gel (SiO_2,Al_2O_3) mixed with zirconium oxide is stationary and kept in a chamber, heated to 400-500ºC. Heavy oil is passed through it at 1.5 kg/cm² pressure. Then it is allowed to pass through fractionating column. The heavy condense at the bottom and recycled. The gasoline fraction is collected at the middle. The uncondensed gas escapes through the outlet.

Comparison of Fixed Bed and Fluid Bed Catalytic Cracking

Property	Fixed Bed Cracking	Fluid Bed Cracking
Catalyst	Kept in a chamber	Circulated with stream
Temperature	425-450°C	About 600°C
Design Cost	Less	More
Yield	Low	High
Catalytic efficiency	Low (Out dated)	High (Modern use)

Knocking

In internal combustion engines mixture of air and gasoline is ignited using spark plug. The flame produced spread rapidly and smoothly.

In certain gasoline the chemical constituents will not spread the flame uniformly. The burning of the fuel increases gradually leading to explosive noise known as 'knocking'.

Knocking reduces the fuel efficiency, poor emission characteristics and environmental concerns.

Knocking depends upon the chemical structure of constituents in the gasoline. The knocking decreases in the following order.

Linear hydrocarbons > Branched hydrocarbons > Cycloalkanes > Olefins > Aromatic hydrocarbons

Anti-knock Agents

The chemicals added with gasoline to reduce the knock are known as anti-knock agents. Tetraethyl Lead (TEL) or Diethyl Telluride is added (0.5 to 2 ppm) for the purpose.

It is believed that free radicals are produced from TEL. It in turn reduces the explosive reaction and thus prevents knocking.

TEL after combustion reduced to metallic lead. It deposits on spark plug and also on engine walls. In addition emitted along with exhaust and hence causing environmental pollution.

Ethylene dibromide is added along with gasoline when TEL is used. It converts the lead formed to volatile lead bromide. So engine is protected from lead contamination.

Indeed lead pollution occurs to environment. Due this reason addition of TEL to gasoline is legally prohibited in recent times.

Octane Number

It is a number to measure the knocking character of a fuel and thus a quality rating.

$$H_3C-(CH_2)_6-CH_3 \quad \text{n-Octane}$$

$$H_3C-\underset{\underset{CH_3}{|}}{\overset{\overset{CH_3}{|}}{C}}-CH_2-\underset{\overset{CH_3}{|}}{CH}-CH_3 \quad \text{Isooctane}$$

n-Octane knocks badly and hence its anti-knock character is considered as zero. Isooctane knocks poorly and it anti-knock value is considered as 100. Since the standards used are octanes, the term octane number is in existence. A fuel with octane number 60 indicates its knocking behaviour same as of a mixture of 60:40 isoocatane and n-octane.

Cetane Number

It is the number assigned to diesel to it s quality. In diesel engines linear hydrocarbons have maximum efficiency. It is found that the ignition characteristic of n-hexadaecane (cetane) is highest and thus taken as 100. In contrast 2-methyl naphthalene is lowest and taken as zero.

$$H_3C-(CH_2)_{14}-CH_3$$

n-Hexadecane (Cetane)

2-Methyl naphthalene

A diesel sample having cetane number 80 means its ignition behaviour similar to 80:20 cetane and 2-methyl naphthalene.

Cetane number decreases in the following order.

n-alkane > cycloalkane > alkene > branched alkane > aromatics.

Diesel Index (DI)

It is the American standard of quantifying the quality of diesel. It is calculated as follows.

$$DI = \frac{\text{Specific Gravity by API x Aniline Point } (^\circ F)}{100}$$

API stands for American Petroleum Institute.

Aniline Point is the minimum temperature at which equal volume of diesel and aniline forms a homogeneous solution.

Higher the diesel index better is the fuel. In general DI = Cetane number x 3.

Synthetic Petrol

Synthetic fuel or synfuel is any liquid fuel obtained from coal, natural gas, or biomass. It can sometimes refer to fuels derived from other solids such as oil shale, tar sand, waste plastics, or from the fermentation of biomatter.

Fischer Tropsch Process

Franz Fischer and Hans Tropsch in the 1920s. The Fischer-Tropsch process transforms gas derived from coal (or other substances) into liquid gas.

It is a catalyzed chemical reaction in which synthesis gas (syngas), a mixture of carbon monoxide and hydrogen, is converted into liquid hydrocarbons of various forms. The most common catalysts are based on iron and cobalt, although nickel and ruthenium have also been used.

$$(2n+1)H_2 + nCO \longrightarrow C_nH_{(2n+2)} + nH_2O$$

where 'n' is a positive integer.

Process conditions and catalyst composition are usually chosen to favor higher order reactions (n>1) and thus minimize methane formation. Most of the alkanes produced tend to be straight-chained, although some branched alkanes are also formed.

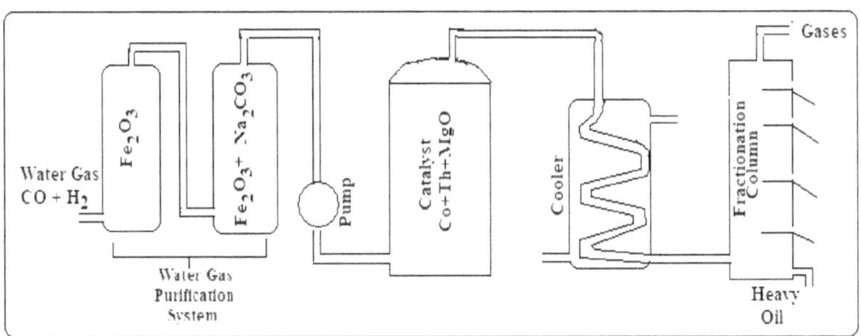

The temperature range of 150-300°C (302-572°F) and increasing the pressure leads to higher conversion rates and also favors formation of long-chained alkanes both of which are desirable. The Fischer-Tropsch catalysts are sensitive to the presence of sulfur containing compounds among other poisons.

Bergius Process

In this process, powdered coal is mixed with heavy oil and heated with hydrogen under high pressure (200-250 atm) at about 748 °K in presence of iron oxide as catalyst.

$$\underset{\text{from coal}}{nC} + (n+1)\,H_{2(g)} \xrightarrow[\text{200 - 250 atm, 748K}]{\text{Iron oxide}} \underset{\text{(mixture of hydrocarbons)}}{C_nH_{2n+2}}$$

The vapours on condensation give a liquid resembling crude oil. This is called synthetic petroleum, which on fractional distillation gives petrol (gasoline).

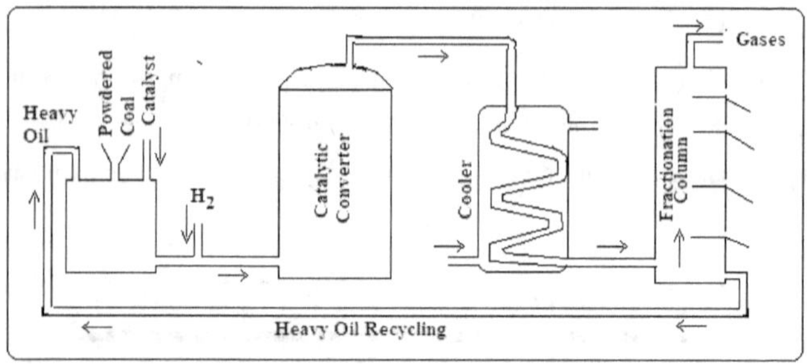

Advantages and Disadvantages Liquid Fuels

Sl.No.	Advantages	Disadvantages
1	High calorific value	Risk of fire hazards
2	Burns without forming ash, clinker, etc.	Give bad odour
3	Fires easily and extinguished easily by stopping liquid flow	Support accessories like tubes, regulators required for controlled use
4	Easy to transport through pipes	Special storage tanks required
5	Clean and economic in use.	

Gaseous Fuels-Natural Gas

It is associated with petroleum deposits. It is mainly methane (lean or dry gas). If it contains propane, butane and other liquid hydrocarbons like pentane, hexane, etc. called as rich or wet gas. Calorific value of natural gas is 12,000-14,000 Kcal/m³. It is used synthetic processes and manufacture of carbon black.

Name of Fuel	Description	Composition (%)	Calorific Value (kcal/m³)	Uses
Water Gas	Passing alternatively steam and little air through a bed of coal or coke maintained at 900-1000ºC	H = 51 CO = 41 CO₂ = 4 N₂ = 4	2,800	Industrial heating, Metallurgy
Producer Gas	Passing air and little steam over coke or coal bed at 1100ºC.	H₂ = 8-12 CO = 22-30 CO₂ = 3 N₂ = 52-55	1,300	Domestic use, Industrial heating
CNG	Natural gas stored in cylinders at high pressure	Methane (<95) Ethane, Propane		Domestic use, fertilizer Manufacture, Source for carbon
LPG	Cracking of petroleum and of heavy oil	Propane 25%, Butane 38, Isobutane37%	27,800	Domestic fuel, vehicle fuel.

Advantages and Disadvantages Gaseous Fuels

Sl.No.	Advantages	Disadvantages
1	Stored large volumes in compressed form	Explosion and uncontrollable
2	Distributed to wide area through pipes	Special storage tanks required.
3	High calorific value	
4	Smokeless and no residue after burning	
5	Easy to transport	

Comparison of Solid, liquid, Gaseous Fuels

Property	Solid	Liquid	Gas
Availability and cost.	Available and cheap	Available in certain countries.	Costly except natural gas
Emissions	High	High	Low
Storage and transport	Stored and transported easily.	Done with risk. Can be transpoted in pipes.	Stored in compressed form, can be transported in pipes.
Ash	Produced	Not produced	Not produced
Calorific Value	Less	Higher	Highest

Flue Gas Analysis

Gaseous emissions containing gases like CO_2, CO, O_2 are called as flue gases. The analysis of flue gases in emissions indicated the pollution potential to the environment and also combustion status like complete or incomplete combustion.

Orsat Apparatus

Exactly 100 ml of flue gas is collected in the burette attached with Orsat apparatus by suitably positioning the water reservoir. The stopper for KOH tube is opened and the gas in the burette is allowed to pass through it for several times. The decrease in volume is read from burette. It corresponds to the volume of CO_2.

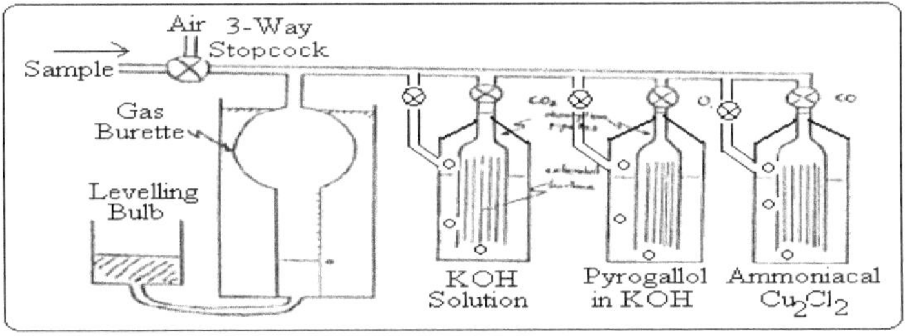

Then the gas is passed through the tube containing alkaline pyrogallol. Oxygen absorbs in this solution. The decrease in volume in the burette is the volume of oxygen in the flue gas mixture.

The contribution of CO is measured by passing through bubble containing ammoniacal cuprous chloride.

The volume of gas remaining after CO_2, CO and O_2 is usually taken as the amount of nitrogen in the flue gas.

Theoretical Air for Combustion

The combustible elements mainly present in a fuel are C, H, S and O. Indeed N, CO_2 and water are non-combustible.

Theoretical air required for the combustion is calculated based on the stoichiometry of the balanced equation for combustion.

Calculation of Air Fuel Ratio

$$\underset{12}{C} + \underset{32}{O_2} \longrightarrow \underset{44}{CO_2}\uparrow$$

(i) Combustion of Carbon

$$\underset{4}{2H_2} + \underset{32}{O_2} \longrightarrow \underset{36}{2H_2O}$$

The weight of O_2 required $= \dfrac{32}{12} \times C$

(ii) Combustion of Hydrogen

$$\underset{28}{CO} + \underset{16}{0.5O_2} \longrightarrow \underset{44}{CO_2}$$

The weight of H_2 required $= \dfrac{32}{4} \times H$

(iii) Combustion of Carbon Monoxide

The weight of CO required $= \dfrac{32}{4} \times CO$

One volume of CO requires 0.5 volume of oxygen.

(iv) Combustion of Sulphur

$$S + O_2 \longrightarrow SO_2$$
$$32 \quad 32 \qquad\qquad 64$$

The weight of S required $= \dfrac{32}{32} \times S$

One volume of S requires one volume of oxygen.

(v) Combustion of Methane

$$CH_4 + 2O_2 \longrightarrow CO_2 + 2H_2O$$
$$16 \quad 64 \qquad\qquad 44 \quad 36$$

The weight of O_2 required $= \dfrac{64}{16} \times CH_4$

One volume of CH_4 requires 4 volumes of oxygen.

(vi) At STP, 1 mole of gas occupies a volume of 22.4 litres and contains Avogadro number of atoms.

Air contains 21% of oxygen by volume and 23 % by weight.

$$\text{For 1 } m^3 \text{ oxygen} = \left(\dfrac{1 \times 100}{23} \right)$$

$$= 4.76 \text{ m}^3 \text{ of air}$$

In terms of weight,

$$\text{For 1 kg oxygen} = \left(\dfrac{1 \times 100}{21} \right)$$

$$= 4.35 \text{ kg of air}$$

Molecular weight of air $= 28.94 \text{ mol}^{-1}$.

(vii) Minimum oxygen required for complete burning = (Theoretical O_2 required – O_2 present in the fuel)

(viii) When both O_2 and H_2 are present in the fuel part of the oxygen is used for the formation of water, which is non-combustible.

$$2H_2 + O_2 \longrightarrow 2H_2O$$
$$4 \quad 32 \qquad\qquad 36$$

Since 32 g of oxygen combines with 4 g of H_2, for 8 parts of O_2 for 1 part of H_2. Thus,

$$\text{Available oxygen} = \text{Mass of } H_2 - \left(\frac{\text{Mass of } H_2}{8}\right)$$

Accordingly, theoretical amount of O_2 required for complete combustion of 1 kg of fuel can be calculated as follows.

$$= \frac{32}{12} \times C + \left(H - \frac{O}{8}\right) + S \text{ kg}$$

(ix) The minimum weight of air required for complete combustion is

$$= \frac{100}{23}\left[\frac{32}{12} \times C + \left(H - \frac{O}{8}\right) + S\right] \text{ kg}$$

(ix) The minimum volume of air required for complete combustion is

$$= \frac{100}{21}\left[\frac{32}{12} \times C + \left(H - \frac{O}{8}\right) + S\right] \text{ m}^3$$

C, H and S stands for percentage of respective atoms in the fuel.

3. Calculate the volume of air required for the complete combustion of 1 litre of CO.

$$CO + 0.5O_2 \longrightarrow CO_2$$

For 1 litre of CO the volume of O_2 require is

$$= \frac{0.5}{21} \times 100 = 2.38 \text{ litres}$$

4. A producer gas sample has the following composition by volume. CH_4 = 3.5%, CO = 25%, H_2 = 10 %, CO_2 = 10.8 %, N_2 = 50. 7%. Calcualte the theoretical amount of air required per cubic meter for complete combustion.

$$CH_4 + 2O_2 \longrightarrow CO_2 + 2H_2O$$
$$CO + 0.5O_2 \longrightarrow CO_2$$
$$2H_2 + O_2 \longrightarrow 2H_2O$$

One m³ of the gas contains 0.035 g of methane, 0.25 g of CO and 0.1 g of H_2.

Thus O_2 required = (2 x 0.035) + (0.25 x 0.5) + (0.1 x 0.5)

$$= 0.245 \text{ m3}$$

$$\text{Air required} = \frac{100}{21} \times 0.245 = 1.17 \text{ m}^3$$

5. Calculate the mass and volume of air required needed for the combustion of 1 kg of carbon.

12 g of Carbon combines with 32 g of O_2

$$C + O_2 \longrightarrow CO_2 \uparrow$$
$$12 \quad 32 \qquad\qquad\qquad 44$$

So 1 kg carbon combines with $\dfrac{32}{12}$ kg of O_2

In terms of air it is

$$= \dfrac{32}{12} \times \dfrac{100}{23} = 11.596 \text{ kg of air}$$

11.596 kg of air occupies a volume of $\dfrac{22400}{28.98} \times 11.596$

$$= 8975.48 \text{ litres} = 8.975 \text{ m3}$$

6. An oil contains 80 % C and 20 % H_2 by weight. Determine the weight of air required per kg of oil if 20 % excess air is used for complete combustion.

$$C + O_2 \longrightarrow CO_2 \uparrow$$
$$12 \quad 32 \qquad\qquad 44$$
$$H_2 + 0.5O_2 \longrightarrow H_2O$$
$$2 \quad\;\; 16 \qquad\qquad 18$$

$$O_2 \text{ required per kg} = \left(\dfrac{32}{12} \times 0.8\right) + \left(\dfrac{16}{2} \times 0.2\right)$$

$$\text{In terms of air the oil requires} = \dfrac{100}{23}\left(2.5 + 1.6\right)$$

$$= 17.83 \text{ kg}$$

$$\text{When 20 \% excess air is used it is} = \dfrac{120}{100} \times 17.83$$

$$= 21.4 \text{ kg.}$$

Assignment Questions

1. Study the relative significance of Proximate and Ultimate analysis of fuels.

2. A fuel has the following analytical data. C= 90%, H = 6%, S = 2.5%, O = 0.5 %, ash = 1%.

   (a) Calculate the quantity of air required for complete combustion of 5 kg of the fuel.

   (b) Perform the calculation if 25% excess air is used for combustion.

3.   Calculate gross and net calorific values of the sample data, C = 80%, H = 7%, O = 3%, S = 3.5%, N = 2.5% and ash = 4%.

4.   Calculate the volume of air required for complete combustion of 2 m³ fuel containing methane (70%) and ethane (30%).

5.   In an Orsat analysis 75 ml of gas decreased its volume 25 ml, 15 ml and 20 ml respectively when passed through KOH, pyrogallol and ammoniocal cuprous chloride solutions. Perform a flue gas analysis.

Self Test

1.   Define calorific value and correlate with heat content.

2.   Analyze the relative merits of solid, liquid and gaseous fuels.

3.   What is metallurgical coke and give its functions?

4.   What is fractional distillation? Apply fractional distillation process to petroleum.

5.   Explain the knocking characteristics and its control methods of fuels.

6.   Explain the design and working principle of Orsat apparatus.

7.   What is octane rating and give its significance?

8.   Analyze the relative merits and demerits of Bergius and Fischer-Tropsch precesses.

9.   Give the composition and use of water gas, producer gas, CNG and LPG.

10.   Recommend strategic methods to reduce flue gas emissions from industries.

PHASE RULE AND ALLOYS

Phase rule, statement and explanation of terms, one component system-water system, condensed phase rule-construction of phase diagram by thermal analysis, simple eutectic system (silver-lead system only), alloys, ferrous alloys, importance, ferrous alloys, Nichrome and stainless steel, heat treatment of steel, non-ferrous alloys, brass and bronze.

Gibbs Phase Rule

The characteristics of heterogeneous system in equilibrium can be studied by theoretical means. J. William Gibbs based on thermodynamic considerations derived an equation called Gibbs rule.

Phase is a qualitative method of a system in identifying its physical characteristics and chemical composition when in equilibrium. It gives a fair idea about the type of equilibrium phenomenon existing between phases.

Statement of Phase Rule

Gibbs Phase Rule states that every heterogeneous system is in equilibrium, the sum of the number of phases (P) and degrees of freedom (F) is greater than the number of components by 2.

$$P + F = C + 2$$
$$F = C - P + 2$$

The rule is valid when there is no external influence like electrical or magnetic interference, surface decomposition, dissociation of molecules, etc. Further, there shall be a definite equilibrium phenomenon existing between phases.

Explanation of Term in Phase Rule

(i) Phase

A phase is a homogeneous physically distinct and mechanically separable part of a system, which can be separated from the rest of the system by definite boundary surfaces.

A homogeneous solution forms one phase. Completely miscible liquids form a single phase. Hydrogen and oxygen in a single container constitutes a single phase.

Mixture of two crystal forms constitutes two phases. Two immiscible liquids are biphasic. Water in a closed container is a two phase system one each for liquid water and water vapour.

(ii) Component

It is the minimum number of independent constituents by means of which the composition of each phase in a system can be explained directly or by means of chemical equation.

$$CaCO_3 \longrightarrow CaO + CO_2$$

It is a triphasic two component system as explained below.

Describing the system in terms of two components CaO and CO_2.

Sl. No.	Phase	Component 1	Component 2
1	$CaCO_3$	CaO	$+ CO_2$
2	CaO	CaO	$+ 0\,CO_2$
3	CO	$0CaO$	$+ CO_2$

Describing the system in terms of two components $CaCO_3$ and CO_2

Sl. No.	Phase	Component 1	Component 2
1	$CaCO_3$	$CaCO_3$	$+ 0\,CO_2$
2	CaO	$CaCO_3$	$- CO_2$
3	CO_2	$0\,CaCO_3$	$+ CO_2$

Liquid water, vapour and ice forms a single component H_2O.

Four phases of sulphur namely, rhombic, monoclinic, liquid and vapour forms single component.

(iii) Degrees of Freedom or Variance

It is the minimum number of independent variables such as temperature, pressure and concentration of a phase to be specified in order to define the system completely.

If a liquid and its vapour are in equilibrim, its component is one, phase is two and its degrees of freedom are one (univariant system). Among three variables temperature, pressure and composition atleast one variable has to be fixed (known) in order to know the other variables of the system.

$$F = 1 - 2 + 2 = 1 \text{ (univariant system)}$$

For a gaseous mixture, phase is one, component is one and degrees of freedom is two (bivariant system). It means by knowing temperature and pressure (independent variables), volume (dependent variable) can be calculated.

$$F = 1 - 1 + 2 = 2 \text{ (bivariant system)}$$

If three phases of a single component is in equilibrium its degrees of freedom is zero (invariant system). This is because all the three phases coexist only at a particular temperature and pressure. So need not fix any variable to define the equilibrium.

$$F = 1 - 3 + 2 = 0 \text{ (invariant system)}$$

In general, for system of given number of components where the number of phases is maximum the degrees of freedom is zero. For a system with given number of phases, larger the number of components greater will be the degrees of freedom. Thus in water if small amount of salt is added the number of degrees of freedom increases from two to three.

Derivation of Phase Rule

Consider a heterogeneous system in equilibrium. It has C components distributed over P phases. At equilibrium temperature and pressure are taken as variables.

Step	Description	Equation
1	Thus the number of variables to be fixed to explain the composition of a phase	= (C – 1).
2	If the system has P phases then total Number of variables required	= P (C – 1)
3	To define the entire system it is necessary to include both the variables namely temperature and pressure also	=P(C-1)+ 2.
4	According to thermodynamics to describe existence of equilibrium between P phases it is required to write P-1 equations.	C(P-1).

The difference of steps three and four gives degrees of freedom (F).

$$F = P(C\text{-}1) + 2 - C(P\text{-}1)$$
$$F = PC - P + 2 - (PC - C)$$
$$F = PC - P + 2 - PC + C$$
$$F = C - P + 2.$$

The above equation is known as phase equation.

Lever Rule

In the phase diagram of solid solutions the composition can be calculated using Lever rule. In solid-liquid equilibrium,

$$\frac{\text{Weight of solid solution}}{\text{Weight of liquid solution}} = \frac{\text{Length of liquid line}}{\text{Length of solid line}}$$

Using lever rule if three quantities are known one can calculate the fourth quantity. Similarly the above equation can be manipulated algebraically.

Phase Diagram

It is a graphical representation of a system in equilibrium. It is drawn using any two parameters among pressure, temperature and composition. If pressure and temperature are used for drawing graph it is PT diagram. On the other hand pressure and composition are used it is PC diagram.

One Component System

It is the study of behavior of one component with respect to temperature and pressure.

Characteristic of One Component System

- The component C = 1. So the phase rule becomes F = 1-P+2 = 3-P.
- Solid-vapour or liquid-vapour equilibrium is always positive for al substances. The line indicating the above will have positive slopes.
- The slope of solid-vapour equilibrium is larger than the liquid-vapour equilibrium. So the line of former is steeper than the latter.
- For solid-liquid equilibrium, the slope is positive except water where it is negative.

Water System

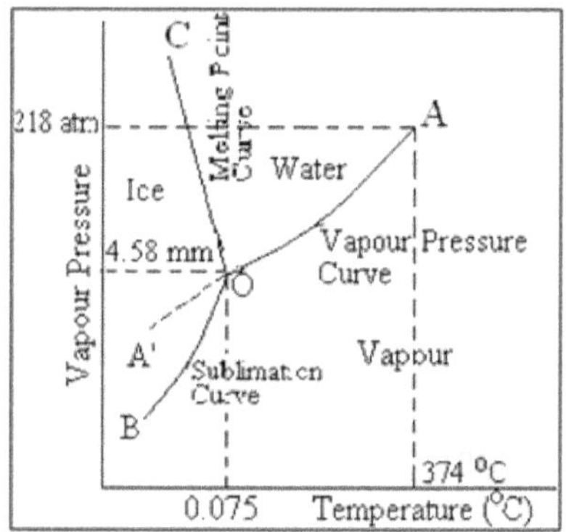

The component water (H_2O) can exist in three different phases namely liquid water, solid ice and water vapour.

Melting point curve of fusion curve for water has negative slope. It indicates melting point of ice decreases with increase in pressure. In other wards, ice melts with decrease in volume. Thus the fact is density of ice is lower than water.

Metastable Equilibrium

Water on cooling it follows the curve OA. On sudden cooling at OA the temperature decreases below zero without forming ice along the curve OA'. The phenomenon of liquid water existing below 0ºC is termed us metastable equilibrium. Thus water below 0ºC is a super cooled liquid. Metastable equilibrium has higher vapour pressure than the ice-water equilibrium. Metastable equilibrium is also observed in other phase diagrams, eg Sulphur.

Sl. No.	Area or curve or point	Phases (P)	Degrees of Freedom
1	Area BOC	Solid ice	2 (Bivarient)
2	Area COA	Liquid water	2 (Bivarient)
3	Area BOA	Water vapour	2 (Bivarient)
4	Curve OB	Ice, vapour	1 (Univarient)
5	Curve OC	Ice, water	1 (Univarient)
6	Curve OA	Water, vapour	1 (Univarient)
7	Curve OA'	Water, vapour	1 (Univarient)
8	Point O	Water, ice vapour	0 (Invarient)

Triple Phase Point

The point O is a triple point. At this point three phases namely, ice, water and vapour are in equilibrium. It is an invariant point because all three phases exist at a particular temperature and pressure. There is no need to mention any variable to describe the state of the system.

Two Component System

When a system is having more than one component the composition of each component has to be specified in order to describe a state of a system completely.

So, the phase diagram has three coordinates namely, temperature, composition and pressure.

Condensed Phase Rule

It is an attempt to simplify the phase diagram. In solid-liquid equilibrium gas phase is neglected. Now the phase rule becomes

$$F = C - P + 1$$

The above equation is known as condensed or reduced phase rule. While applying condensed phase rule vapour phase is ignored. In other wards the study is carried out in an open system.

Construction of Phase Diagram by Thermal Analysis

Pure element or mixture of elements on cooling gives solid at a definite temperature. A graph of temperature with respect to time is called cooling curve.

Recent times thermal analysis is carried out in Digital Scanning Calorimeter (DSC).

On solidification process the temperature remains constant for some time till the process is completed. The horizontal portion of the cooling curve is known as the halt. It is due the heat evolved when the liquid solidifies. During the time period solid liquid equilibrium exists. The concept is used for the construction of phase diagram.

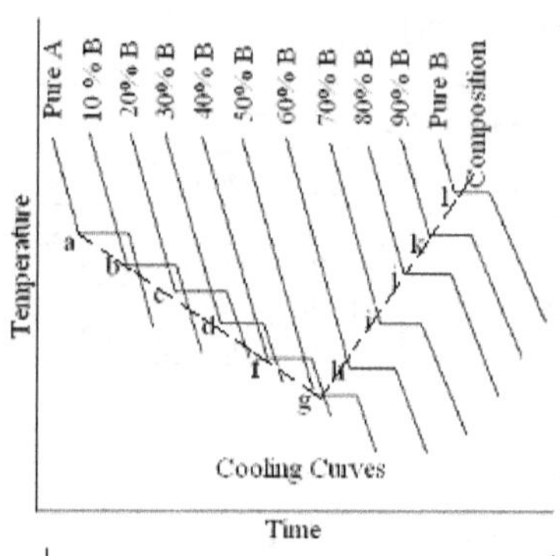

Simple Eutectic System

Eutectic means easy melting mixture. Eutectic mixture is always having lower melting point than its components. Simple indicates the formation of only one eutectic with definite composition. Eg. Lead-Silver system.

Lead melting point = 327ºC

Silver melting point = 961ºC

Eutectic mixture melting point = 303ºC

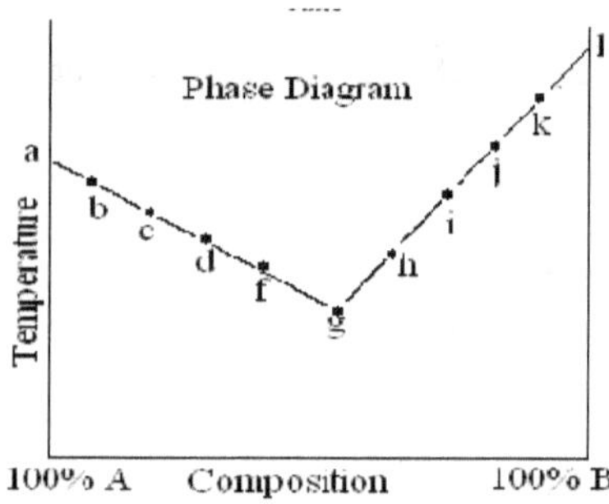

Lead-Silver System (Condensed Phase Diagram)

It is an example for two component system with the formation of simple eutectic mixture.

Eutectic Point Characteristics of Lead-Silver System

Eutectic composition = 97.4 % Pb and 2.6 % Ag

Eutectic Temperature = 303ºC

Eutectic Phases = Solid Ag ⇌ Solid Pb ⇌ Liquid

Maximum amount of silver crystallizes out at eutectic temperature.

Points

S. No.	Point	Phases	Degrees of Freedom
1	A	Pure Ag	1 (Univarient)
2	B	Pure Pb	1 (Univarient)
3	C	Solid Ag + Solid + Liquid	0 (invarient) Eutectic Point

Area

S. No.	Area	Phases	Degrees of Freedom
1	Above ACB	Liquid	2 (Bivarient)
2	ACD	Liquid + solid Ag	1 (Univarient)
3	BCE	Liquid + solid Pb	1 (Univarient)
4	CDFG	Eutectic + solid Ag	1 (Univarient)
5	CGHE	Eutectic + Solid Pb	1 (Univarient)

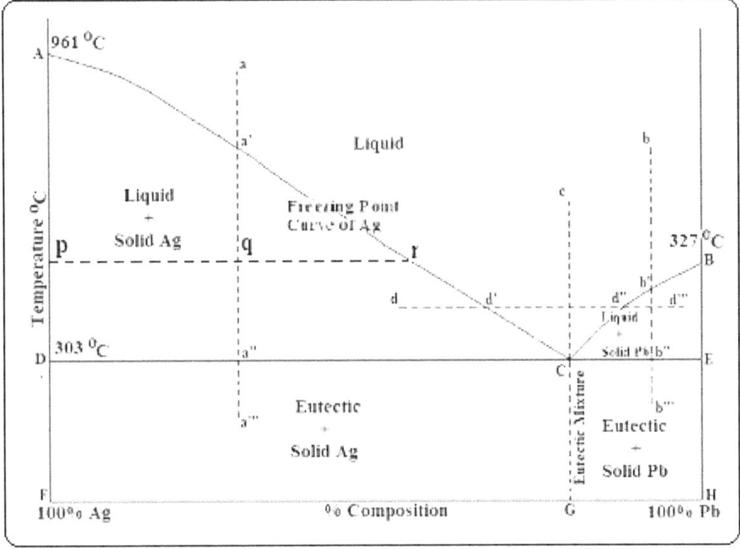

Cooling Curves (Isopleth)

A graph drawn for a system at constant composition is subjected to temperature change is known as isopleth. A composition a, is in liquid state. On cooling, at a' solid silver starts appearing. Along a'a" solid-liquid equilibrium exists. At a" the system becomes completely solid consisting of eutectic mixture and solid silver. If a system at a' is not subjected to external cooling it follows a'C equilibrium. It indicates more and more lead start crystallizing at the expense of temperature of the system. Thus temperature decreases along AD and composition changes along a'C. At C eutectic mixture is formed. Then it follows CG curve.

A similar phenomenon occurs if a system at b is cooled along bb'''. Here solid lead appears first then eutectic mixture is formed.

A system at c on cooling follows cCG. At point C only eutectic mixture is formed as solid phase and continues along CG.

Isobars

A graph of an experiment of subjecting a system at constant temperature with varying composition is known as isobar. A system at d consists of liquid Ag and solid Ag.

On adding silver at constant temperature the composition varies along dd'. At d' the system becomes homogeneous liquid and follows along d'd''. At d'' solid lead starts forming though silver added. Along d''d''' liquid-solid lead phases exist.

Quantification of Components in Solid-Liquid Equilibrium

The quantity of solid and the liquid present in solid-liquid equilibrium can be calculated using lever rule. The quantity of silver at the point q can be calculated as follows.

$$\frac{\text{Mass of solid Ag}}{\text{Mass of liquid}} = \frac{qr}{pq}$$

Significance of Lead-Silver System

(Pattinson's Process of Desilverisation of Argentiferrous Lead)

A homogeneous solution of liquid containing less than 2.6 % silver on cooling, lead separates out leaving silver in solution. On repeating the process the amount of lead decreases and silver gets purified. Pattinson used this concept for the purification of silver.

Similarly, any composition above 2.6 % silver on cooling silver separates out. Thus percentage of lead increases in solution. By repeating the process pure lead can be obtained.

Uses of Phase Diagram (Rule)

- It is independent of amount of substance in the system but only depends on percentage composition.
- It applies to study both physical and chemical equilibrium.
- Formation of compounds and metastable equilibrium can be studied.
- It helps in purification of metals and formation of alloys.

- It predicts the existence of formation of compounds or simple mixtures.
- It is an important generalization and enables to study and classify equilibrium of more complex heterogeneous systems.

Limitation of Phase Diagram (Rule)

There shall be an equilibrium to study phase rule. Thus all the phases should exist in same condition.

Limited to variables under study generally only for temperature, pressure and composition.

The rule does not depend on the nature or amount of substance present in the equilibrium.

The effect of other variables like electric, magnetic, gravitational, surfaces forces, molecular interactions, secondary forces, etc are not taken into consideration.

It is a macroscopic phenomenon based on classical thermodynamics. Microscopic information like molecular structure, motion, etc. can not be inferred.

It is a qualitative prediction based on phases, equilibrium, etc.

Phase rule is not applicable to homogenous equilibria.

Alloys

An alloy is a metallic, homogeneous mixture of metals or elements in which atleast one component is a metal. Alloys are considered as solid solutions.

Importance of Alloys

Alloys improve physical and chemical properties. Better corrosion resistance, hardness, thermal stability, colour, moulding property, etc are some of the improvements achieved by alloying. Indeed super- and semi-conductors are alloys.

Specific examples include,

- Bronze an alloy of copper is more corrosion resistant than copper.
- Brass an alloy of copper and zinc is yellow in colour.
- Wood's alloy has a melting point of $71^{\circ}C$ which is lower than any of its components (Bi, Pb, Cd and Zn)

Ferrous Alloys

Alloys containing iron as one of the major components is called ferrous alloys. Eg nichrome, steel, etc.

Nichrome

It is a nickel–chromium alloy with iron and manganese as other elements.

Composition

Ni = 60 %, Cr = 12 %, Fe = 26 % and Mn = 2 %.

Properties

- Resistant and oxidizing agents.
- Poor conductor of electricity.
- Stable to heat upto 900ºC.

Uses

- Manufacture of heating elements and resistance coils.
- Manufacture of gas turbines, valves and parts for boilers.
- Making high temperature stable machineries.
- Used in making elements for electric iron.

Stainless Steel

Stainless steel is a class of alloys of iron with elements like Cr, Ni, Mo and C.

Heat Treatment of Steel

Heat treatment is the controlled conditioning of steel by heating followed by cooling. The heating rate, maximum temperature, cooling rate, etc. decides the quality of steel formed. A typical heat treatment is selected based on the application of steel. On heat treatment composition remains constant but changes in physical properties occurs.

Uses of Heat Treatment

- Improves microstructure and mechanical and processing qualities.
- Relieves internal strain.
- Changes in electrical and magnetic properties.
- Improves hardness, ductility and tensile strength.

1. Normalising

In this method steel containing less than 0.8 % C (Hypoeutectoid steel) is heated to 950ºC. If it has carbon more than 0.8%C (Hypereutectoid steel) heated to above 1000ºC. Maintained for some time and cooled fast by air.

- Normalizing improves fine structure of steel.
- It forms alternate ferrite (practically pure iron) and cementite (Fe_3C) structure.

2. Annealing

It is the process of heating steel to a predetermined temperature, holding it for some time (socking) followed by cooling at a very slow rate.

Methods of Annealing

Name of Annealing	Condition	Purpose
Full Annealing	Heated to 910 ºC for low carbon steel (0.8% C) or 725 ºC for high carbon steel (2% C). Cooled slowly.	Improves hardness, machinability. Removes internal stress.
Spheroidising	Maintained just below the critical temperature for long time and then cooled.	Increases softness and machinability. Forms sphere shaped cementite (Fe_3C) in ferric matrix (spheordite).
Diffusion Annealing	Kept for 10-20 h at 1000-1200ºC. Cooled slowly.	Attains homogeneous composition. Improves hardness.
Recrystallisa-tion Annealing	Heated to 675ºC and cooled slowly.	Formation of ferrite. Essential for heavy cold working of steel.
Partial Annealing	Slow heating upto 725ºC then cooled slowly.	Improve flexibility and machinability.
Stress Relief Annealing	At 500-525ºC for 1-2 h and then cooled slowly.	Relives residual stresses.
Isothermal Annealing	Heated upto 725ºC maintained for some time. Cooled fast using air.	Homogeneous structure obtained. Improves machinability.

Difference between Annealing and Normalizing

S. No.	Annealing	Normalizing
1	Cooling done in furnace	Cooled in air
2	Cooled at a rate of 1ºC/sec.	Cooled at a rate of 20ºC/sec.
3	Coarse pearlite (ferrite + cementite) microstructure.	Fine pearlite microstructure.
4	Uniform distribution of grain size.	Less uniform distribution of grain size.
5	Internal stresses are almost completely removed.	Not fully removed.
6	More suitable to improve machinability of low carbon steel.	More suitable to improve machinability of high carbon steel.
7	Comparatively attains less hardness, toughness and tensile strength.	Comparatively attains more hardness, toughness and tensile strength.

3. Hardening (Quenching)

Steel is heated to above 1000 ºC and cooled adiabatically by dipping in cold oil of brine water.

Low carbon steel can not be used for this purpose since it is soft and brittle.

- Hardening improves abrasion resistance.
- It increases the cutting strength and resistance to wear and tear.

Tempering

It is the heat treatment to make homogeneous structure, improve ductility and reduce the internal stress.

Methods of Tempering

(i) Low Temperature Tempering

Heated to 150-200ºC and hold for some time then cooled slowly. It relieves internal stress and improves ductility. It is done for steel used for making cutting tools and machine part.

(ii) Medium Temperature Tempering

Heated between 350-450ºC and cooled slowly. It is the process for steel when used for making springs, plates, etc.

(iii) High Temperature Tempering

Steel is heated to 500-600ºC and cooled to room temperature. It removes internal stress completely. Steel used for high impact machine parts like gear wheels, shafts, connecting rods, etc. undergo this treatment.

- Tempering removes internal stress and strain if exists.
- It provides ductility with toughness.
- Tempering provides required qualities for cutting tools.

5. Carburising

Steel is heated and maintained at 900-950ºC in the presence of carbon and cooled slowly to room temperature. Surface penetration of carbon occurs and surface hardness is obtained. It leads to the formation of hypereutectoid steel (0.8-1.2 %C).

- Increases the surface hardness
- Steel becomes wear and tear resistant.

6. Nitriding

Steel is heated to 550ºC in the presence of ammonia atmosphere. Under this condition ammonia decomposes to nitrogen and hydrogen. This nitrogen reacts with iron forms nitride alloy on the surface.

- Gives super hard steel
- Enhances the corrosion resistance.

7. Cyniding

Heating to 800-900ºC is done in the presence of sodium cyanide and quenched in oil. It allows the formation of hard cyanide layer over steel.

- It imparts surface hardness.
- It gives shiny surface finish and corrosion resistance.

Non-Ferrous Alloys

Alloys which do not contain iron as one of the component is called non-ferrous alloys. They have special properties and applications in material science and engineering. They have qualities like corrosion resistance, softness, special electric and magnetic properties, low coefficient of friction, etc.

Brass

Brass is a class of copper alloy with zinc and with or without other metals.

Name	% Composition			Characteristics	Uses
	Cu	Zn	Sn		
Guiding metal brass	90	10		Golden colour Strong and hard	Rivets, screws, jewellery.
Dutch metal brass	80	20		Golden colur. Strong and hard.	Battery caps, tubes, jewellery.
Cartridge brass	70	30		Soft, ductile. Can be cold deformable.	Cartridge cases
Muntz brass	60	40		Hard, corrosion resistance	Volves, crews, marine fittings
Naval brass	62	37	1	Hard, corrosion resistance	Condenser tubes, marine engine parts.
German silver	50	20	30	Good strength Corrosion resistance.	Utensils, screws, decorative articles.

Bronze

It is also an alloy of copper contains 80-95% copper. Other elements are tin, lead, etc. It is tough, workable and corrosion resistance.

Name	%Composition	Characteristics	Uses
Coinage metal	Cu = 89-92 % Sn = 8-11 %	Soft, ductile and durable.	Coins, volves, ornaments.
Gun metal	Cu = 85 % Sn = 5 % Zn = 5 % Pb = 5 %	Hard, tough, resistant to explosion	Heavy load bearing tanks, steam plants
Phosphor bronze	Cu = major Sn = little P = little	Strong, high load bearing	Gears, volves, turbine blades.
Aluminum branze	Cu = 90 % Al = 10 %	Malleable and ductile, hard	Castings, bushes, bearings.
Nickel branze	Cu = 89-92 % Sn = 8-11 %	Corrosion resistant, hard	Hydraulic components, valves
Constantan	Cu = 60 % Ni = 40 %	High electrical resistance	Resistance and thermocouple wires

Assignment Questions

1. For the following equilibria calculate the number of components, phases and degrees of freedom.

    (i) $PCl_{5(g)} \rightleftharpoons PCl_{3(g)} + Cl_{2(g)}$

    (ii) $NaBr(s) \rightleftharpoons Na^+_{(aq)} + Br^-_{(aq)}$

    (iii) $NaCl + KBr \rightleftharpoons NaBr + KCl$

    (iv) $Fe_2O_{3(s)} + CO_{(g)} \rightleftharpoons 2FeO_{(s)} + CO_{2(g)}$

2. Construct a phase diagram for the following data and find out the eutectic point, temperature and composition.

Mass (%B)	0.0	10	25	35	45	55	65	80	92	100
Freezing Point (°C)	119	112	99	90	78	83	95	108	113	115

3. Explore the effect of melting point and conductivity on addition of metallic, non-metallic and metalloid impurities in iron.

4. Comment on the phase diagrams of the following one component equilibrium.

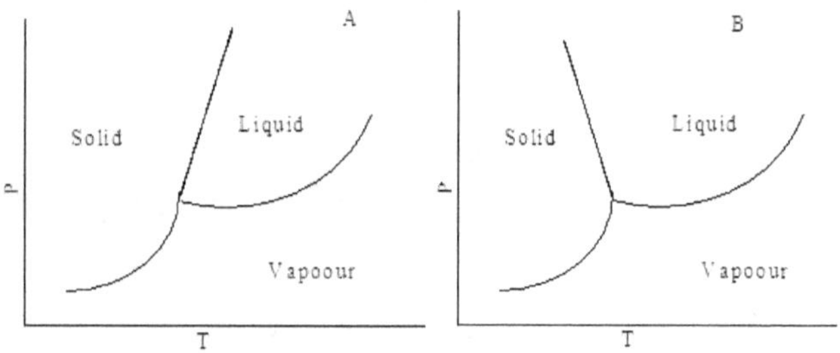

5. Draw a thermal analysis graph for lead-silver system.

Self Test

1. State and explain the terms in Phase rule.
2. Identify the following statements are true or false.

   - For one component system maximum phases that can coexist together is three.
   - A system can have negative degrees of freedom.
   - All open system follows the phase rule $P + F = C + 2$.
   - Study of phase equilibrium gives information regarding structure of matter in the equilibrium.
   - The stable allotropic modification of a substance has a higher melting point than the unstable or less stable form.

3. Distinguish the following

- Stable and unstable equilibrium
- Eutectic point and solid solutions
- Compound and solid solution

4. Solid-liquid equilibrium for water is negative where the same is positive for other substances.

5. Why phase rule is condensed for two component systems.

6. Bring out the uses and limitations of phase rule.

7. Write a note on non-ferrous alloys.

8. What is alloying and the principle to be followed while selecting and alloying element.

9. Explain the phase diagram of lead-silver system and its industrial significance.

10. Analyze the different heat treatment process performed on steel and its merits.

UNIT V

ANALYTICAL TECHNIQUES

Beer-Lamber's law (problem)-UV-visible spectroscopy and IR spectroscopy-principles-instrumentation (problem) (block diagram only)-estimation of iron by colorimetry-flame photometry-principle-instrumentation (block diagram only)-estimation of sodium by flame photometry-atomic absorption spectroscopy-principles-instrumentation (block diagram only)-estimation of nickel by atomic absorption spectroscopy.

Analytical Techniques

Analytical techniques deal with the qualitative and quantitative identification of materials. It includes chemical and physical techniques.

Chemical methods include salt analysis, gravimetry, titrometry, etc.

Physical methods are determination of melting point, optical rotation, spectral studies, etc.

Spectroscopy is a branch of science, which deals with the interaction of electromagnetic radiation with matter. It is based on the principle of quantization of energy levels and dual nature of electromagnetic radiation.

Spectroscopy is used to determine identity, quantity, structure, and the environment of atoms, molecules, and ions by analyzing the radiation emitted or absorbed by atoms, molecules, or ions.

Analytical spectroscopy is a specific study of the use of spectroscopic techniques for qualitative and quantitative analysis. A spectrum is a graphical representation of interaction of electromagnetic radiation with molecules and atoms.

Absorption and Emission Spectroscopy

In absorption spectroscopy, spectrum is produced due to the absorption of incident radiation by molecules.e. g., UV-spectrum, IR spectrum, etc.

In emission spectroscopy, spectrum is produced due to the emission of radiation by molecules.e.g., Raman spectroscopy, flame spectroscopy, fluorescence spectroscopy, etc.

Electromagnetic Radiation

Entire range of radiations available for human perception directly or indirectly constitutes electromagnetic spectrum.

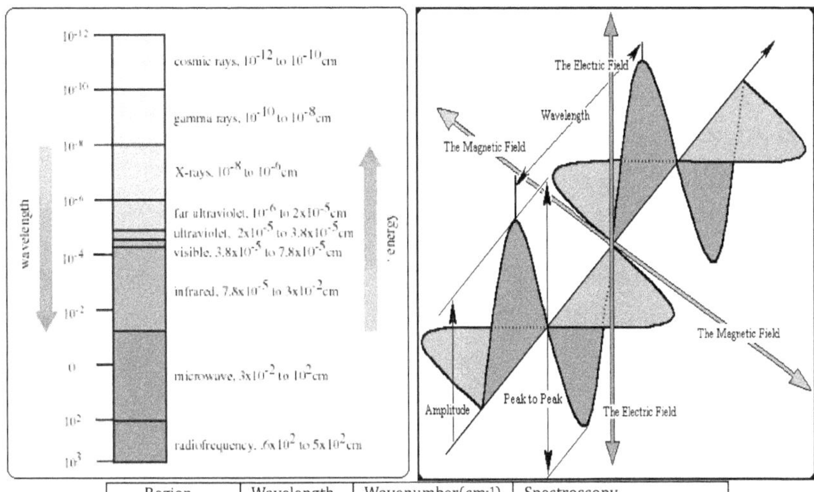

Region	Wavelength	Wavenumber(cm⁻¹)	Spectroscopy
Cosmic Rays	10^{-12}-10^{-10} m	$> 10^{12}$	
γ- Rays	10^{-10}-10^{-8} m	10^{9}-10^{12}	
X-Rays	10^{-2}-10^{2} Aº	10^{6}-10	X-ray Spectroscopy
Far Ultraviolet	10-200 nm	10^{3}-10^{6}	
Near Ultraviolet	200-400 nm	25000-10^{3}	UV-Spectroscopy
Visible	400-750 nm	25000-13000	
Near Infra red	0.5-2.5 µm	13000-4000	
Mid Infra red	2.5-50 µm	4000-200	IR- Spectroscopy
Far Infra red	50-1000 µm	200-10	
Microwaves	0.1-100 cm	10-10^{2}	Rotational Spectroscopy
Radiowaves	1-1000 cm	< 100 µm	EPR and NMR spectroscopy

According to Plank quantum theory electromagnetic radiations are propagated as discrete energy packets in the form of wave.

Energy, E = hυ

h = Plank's constant =6.627 x 10^{-34} Js, υ = frequency of radiation.

Energy of electromagnetic radiation is directly proportional to frequency of radiation.

$$E = h\frac{c}{\lambda} \qquad \left[\because \upsilon = \frac{c}{\lambda} \right]$$

c = speed of light = 3 x 10^{-8} ms⁻¹, λ = wave number.

Energy of electromagnetic radiation is indirectly proportional to wave number.

$$E = hc\bar{\upsilon} \qquad \left[\because \bar{\upsilon} = \frac{1}{\lambda}\right]$$

$\bar{\upsilon}$ = wave number

Thus energy of electromagnetic radiation is directly proportional to wave number of radiation.

Beer-Lambert's Law

For dilute solutions the quantity of light absorbed is directly proportional to the concentration (C) and path length (l) of the absorption medium.

$$\log \frac{I_o}{I} \; \alpha \; Cl$$

I_o = Intensity of incident light; I = Intensity of emitted light.

$$A \; \alpha \; Cl \qquad \because A = \log \frac{I_o}{I}$$

$$A = \varepsilon Cl$$

A= Absorbance; ε = Molar Extinction Coefficient.

When path length is kept constant, the equation becomes,

$$A = \varepsilon C$$

$$\log \frac{100}{90} = \varepsilon \times C_1 \times l \qquad A \, \alpha \, C$$

It means when path length is maintained constant the absorbance of a solution is directly proportional to concentration.

Applications of Beer-Lambert Law

- It is used to determine the unknown concentration of solution.
- Absorption in UV-Spectroscopy is used in identification of electronic transitions.
- Colorimetric estimation of compounds are based on Bear-Lambert's law.
- Reaction kinetics, stability constant measurements, etc are based on the principle of Beer-Lambert's law.

Limitations of Beer-Lambert's Law

- The law is applicable only for dilute solutions.
- It is useful only when monochromatic radiation is used.
- The compound should absorb the radiation
- The temperature of the system should be constant.

- The law fails if there is a molecular association or dissociation.

Problems

1. A monochromatic radiation when passed through 10 cm length solution of concentration 0.05 M decreased one fourth of its initial value. Calculate molar extinction coefficient.

$$\log \frac{Io}{I} = \alpha \, Cl$$

$$Io = 100, \ I = 25.$$

$$\log \frac{100}{25} = \varepsilon \times 0.05 \ \text{moldm}^{-3} \times 10 \ \text{cm}$$

$$\varepsilon = 1.204 \ \text{dm}^3 \ \text{mol}^{-1} \ \text{cm}^{-1}$$

2. A substance at 10-3 M concentration absorbed 10 % of incident radiation. Calculate concentration required for 90 % absorption of the same radiation.

$$\log \frac{Io}{I} = \alpha \, Cl$$

Case I

When concentration is C_1, the solution absorbed 10 % of incident radiation.

$$\log \frac{100}{90} = \varepsilon \times C_1 \times l \qquad \text{(i)}$$

Case II

When concentration is C_2, the solution absorbed 90 % of incident radiation.

$$\log \frac{100}{10} = \varepsilon \times C_2 \times l \qquad \text{(ii)}$$

From equation (i) and (ii),

$$\frac{\log \dfrac{100}{90}}{\log \dfrac{100}{10}} = \frac{C_1}{C_2} = \frac{0.001 \ \text{mol dm}^{-3}}{C_2}$$

$$C_2 = 0.022 \ \text{mol dm}^{-3}$$

3. The molar extinction coefficient of iron (II) phenanthroline complex is 12,000 dm mol^{-1} cm^{-1} and the minimum detectable concentration is 0.01 M. Calculate the minimum molar concentration that can be detected based on Beer-Lambert's law when path length is 1 cm.

$$A = \varepsilon Cl$$

$$C = \frac{A}{\varepsilon l} = \frac{0.01}{12000 \ \text{dm}^3 \ \text{mol}^{-1} \ \text{cm}^{-1} \ 1 \ \text{cm}}$$

$$C = 1.2 \times 10^{-6} \ \text{mol dm}^{-3} = 1.20 \times 10^{-6} \ \text{M}$$

UV-Visible Spectroscopy

Electromagnetic radiations having wavelength range 100 nm-340 nm constitute ultraviolet region. Visible spectrum is between 340 nm and 760 nm. The wavelength below 200 nm upto 10 nm is considered as vacuum ultraviolet region. Above 760 nm upto 1100 nm is far UV region.

The spectroscopic determination based on the use of ultraviolet and visible light is called ultraviolet spectroscopy. It is an absorption spectroscopy.

Principle

Interaction of molecules with ultraviolet and visible radiation causes electronic transition. The wavelength and intensity of absorption is used for identification, qualitative and quantitative analysis.

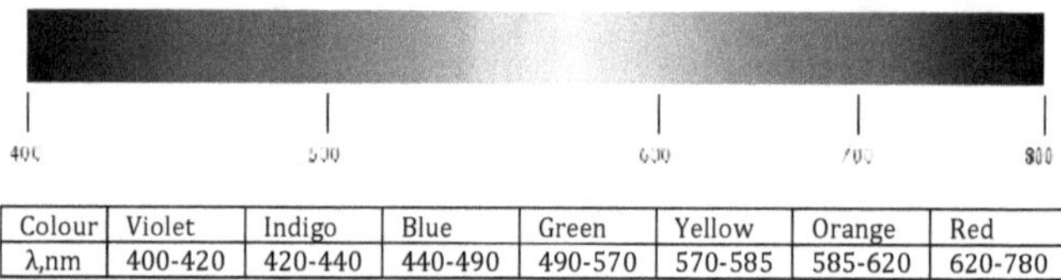

Colour	Violet	Indigo	Blue	Green	Yellow	Orange	Red
λ,nm	400-420	420-440	440-490	490-570	570-585	585-620	620-780

Instrumentation (Block Diagram)

Single Spectrophotometer

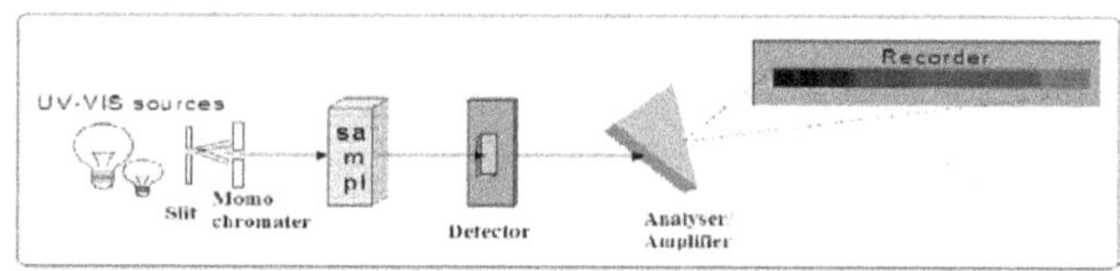

Double Spectrophotometer

A double beam instrument incident beam takes splits into two beams. One is sample beam and a constant reference beam. This allows for internal compensation for any change in light intensity.

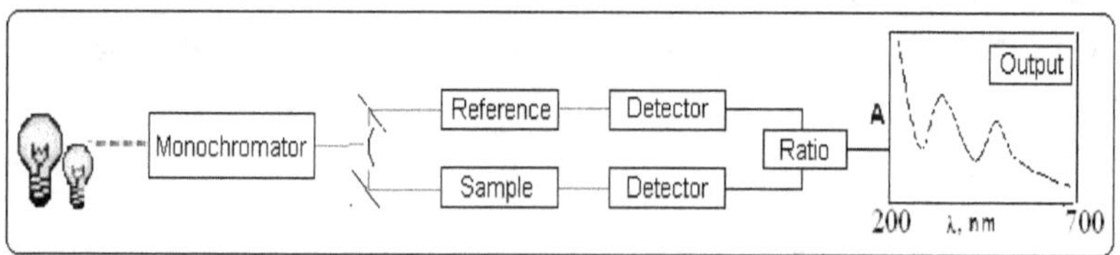

Advantages of Double Beam Spectrophotometer

- Higher optical stability and greater reproducibility.
- Variation in environmental condition of measurement like temperature fluctuation is partly taken care of by double beam spectrophotometers.
- Greater sensitivity to measure low concentrations especially in biological samples can be analyzed.
- Turbid samples can be used in double beam instruments. It is extensively used in cell culture studies.

Light Source

Deuterium lamp covers the UV region 200-340 nm. The mechanism for this involves formation of an excited molecular species, which breaks up to give two atomic species and an ultraviolet photon.

$$D_2 + \text{electrical energy} \longrightarrow D_2^* \longrightarrow D' + D'' + h\nu$$

Tungsten lamp covers region 340-900 nm. The energy emitted by a tungsten filament lamp is proportional to the fourth power of the operating voltage. This means that for the energy output to be stable, the voltage should be stable.

Monochromator

All monochromators contain the following component parts.

- An entrance slit
- A collimating lens
- A dispersing device (usually a prism or a grating)
- A focusing lens
- An exit slit

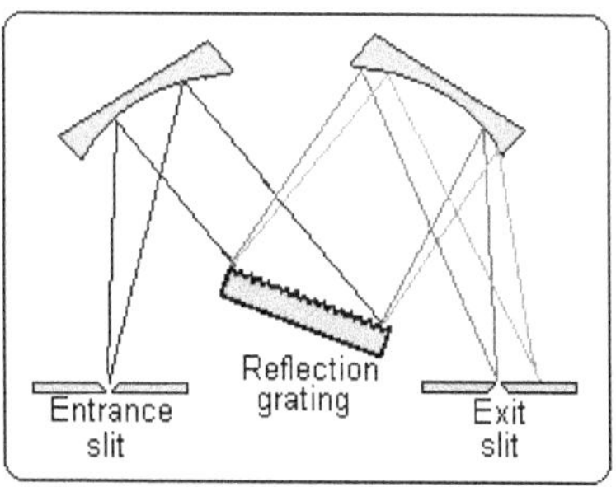

Detectors

- Photomultiplier Tube
- Diode Array Detectors
- Charge Coupled Devices

The photomultiplier tube is a commonly used detector in UV-Visible spectroscopy. It consists of a photoemissive cathode (a cathode which emits electrons when struck by photons of radiation), several dynodes (which emit several electrons for each electron striking them) and an anode.

These electrons are then accelerated towards the second dynode, to produce more electrons which are accelerated towards dynode three and so on. Eventually, the electrons are collected at the anode. By this time, each original photon has produced 10^6 - 10^7 electrons. The resulting current is amplified and measured.

The linear photodiode array is an example of a multichannel photon detector. These detectors are capable of measuring all elements of a beam of dispersed radiation simultaneously.

A linear photodiode array comprises many small silicon photodiodes formed on a single silicon chip. There can be between 64 to 4096 sensor elements on a chip, the most common being 1024 photodiodes.

Charge-Coupled Devices (CCDs) are similar to diode array detectors, but instead of diodes, they consist of an array of photocapacitors.

Frank- Condon Principle

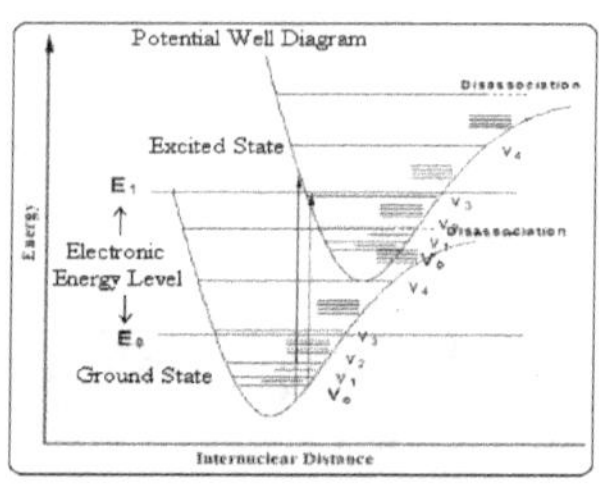

Electronic transition from ground state to excited state is faster than nuclear motion (molecular vibrations). Electronic transitions are in the time scale of 10^{-12} sec. Molecular vibrations are in the time scale 10^{-9} sec.

Frank Condon principle is the consequence of Heisenberg's uncertainty principle.

Applications of Frank Condon Principle

- The concept allowed and forbidden electronic transitions are based on the concept of Franck-Condon Principle.
- Intensity of absorption in UV-spectroscopy is based on Frank-Condon Principle.
- Selection rule is based on Frank-Condon Principle.

Selection Rule

"Transitions may occur only between energy states with the same spin multiplicity"

Electronic Spectra is Band Spectra

Electronic spectra appear as broad bands. This is due to every electronic transition is accompanied by vibrational and rotational transitions. So closed packed lines leading to band spectra.

Relative Energy of Electronic Transition

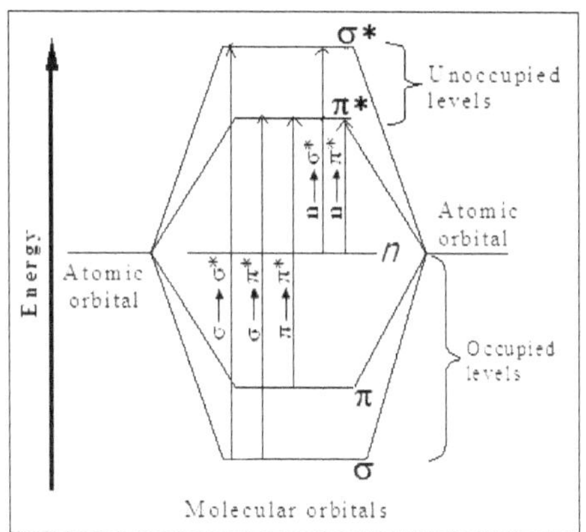

$$\sigma \longrightarrow \sigma^* > \sigma \longrightarrow \pi^* > \pi \longrightarrow \pi^* > n \longrightarrow \sigma^* > n \longrightarrow \pi^*$$

Types of Electronic Transition

Transition	Compound	Absorption Wavelength
$\sigma \longrightarrow \sigma^*$	alkanes	~150 nm
$\sigma \longrightarrow \pi^*$	carbonyls,	~170 nm
$\pi \longrightarrow \pi^*$	unsaturated cmpds	~180 nm
$n \longrightarrow \sigma^*$	O, N, S, halogens	~190 nm
$n \longrightarrow \pi^*$	carbonyls	~300 nm

Chromophores

Light absorbing part of the molecule is called chromophore.

Eg.

Auxochromes

A group or substituent increase the wavelength of absorption is called auxochromes. Eg. Br, OH, NH_2, etc.

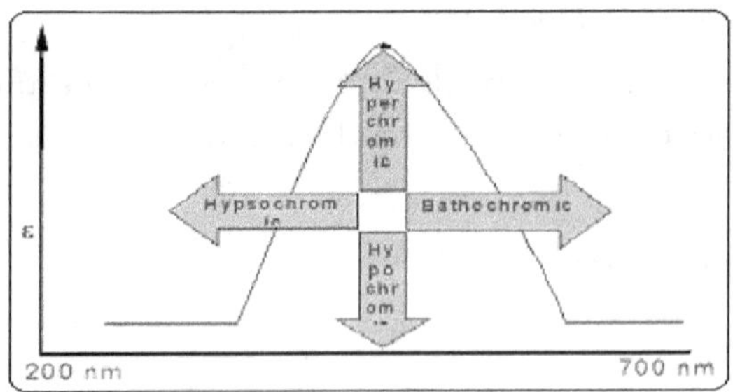

Bathochromic Shift (Red Shift)

It is the shift of absorption spectrum towards longer wavelength.

Hypsochromic Shift (Blue Shift)

It is the shift of absorption spectrum towards shorter wavelength.

Characteristic Features of UV-Spectra

A. Wavelength of Absorption

It gives an idea of structural feature of the molecule. Coloured compounds will absorb in the visible region. It indicates the compound has extended conjugated double or triple bonds.

The presence of many absorption maxima indicated the presence of different type of transitions.

B. Intensity of Absorption

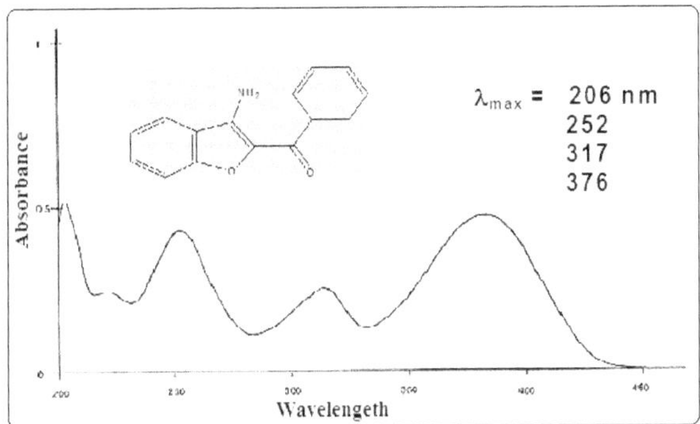

It corresponds to peak height indicative of molar absorption (extinction coefficient). The stronger the band greater the accuracy when used for quantitative estimations.

Based in the intensity of absorption allowed and forbidden transitions can be inferred. In otherworld it gives an inference of molecular symmetry.

- values of 10^4-10^6 are termed high intensity absorptions and spin allowed transitions.
- values of 10^3-10^4 are termed low intensity absorptions and symmetry allowed transitions.
- values of 0 to 10^3 are the absorptions of forbidden transitions.

A is unitless, so the units for ε are $cm^{-1}M^{-1}$ and are rarely expressed.

Theoretical Calculation of Absorption Maxima

Woodward-Fischer Rule

Diene	Transoid	Cisoid
Structure		
Base Value, λmax (nm)	215	254

S. No.	Group	Increment (nm)
1	Extended conjugation	+30
2	Each exo-cyclic C=C	+5
3	Alkyl	+5
4	-OR	+6
5	-SR	+30
6	-Cl, -Br	+5
7	$-NR_2$	+60
8	$-OCOCH_3$	+0

Molecule	Groups	Wavelength (nm)
	Base	215
	Alkyl Group	5
	Total	220 (Calculated)
		222 (Observed)
	Base	215
	Exocylic	5
	2-Alkyl Groups	10
	Total	230 (Calculated)
		235 (Observed)
	Base (Cisoid)	254
	Exocylic	5
	Alkyl Group	5
	Extended Conj.	30
	-N(CH$_3$)$_2$	60
	Total	354 (Calculated)
		350 (Observed)

Determination of Unknown Concentration by UV-Spectroscopy

Principle

The Fe^{2+} ions forms coloured complex with 1,10-Phenanthralein having absorption maximum at 512 nm.

For dilute solutions the absorption is directly proportional to concentration (Beer-Lambert's Law).

A = ε Cl

A = Absorbance

ε = Extinction Coefficient

A standard Fe^{2+} solution is prepared using ferrous ammonium sulphate. From this Fe^{2+} concentrations 5, 10, 15, 20, 25 ppm is prepared.

To 0.5 ml of the above independent solution in test tubes 1 ml each of 10% hydroxylamine and 0.2% sodium acetate solutions are added and mix well. Further, 0.5 ml of 0.25% 1,10-phenanthralein solution are added and mixed well and allowed to stand for 5 minutes. A blank experiment is carried out using 1 ml of distilled water instead of sample. Appropriate amount of distilled water is added in order maintain constant volume in all the experiments. The absorbance at 512 nm against blank is read in UV spectrophotometer and tabulated the

readings (table 1). Blank value is used for zero correction in the case of single beam instrument.

Expt. No.	Solution	Absorbance at 512 nm
1	Blank	
1	5 ppm	
3	10 ppm	
4	15 ppm	
5	20 ppm	
6	25 ppm	
7	Unknown	

A graph is drawn between concentration (x-axis) and absorbance (y-axis). It gives a straight line passing through the origin.

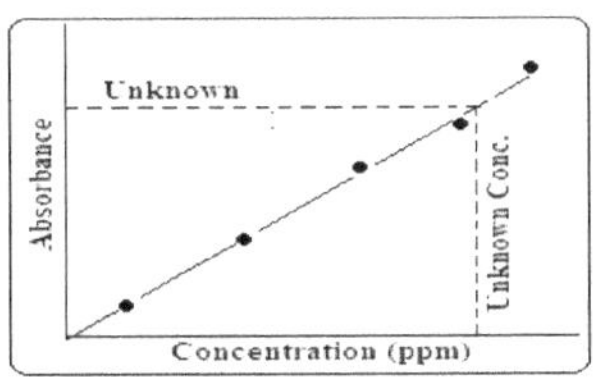

Determination of Unknown Concentration

The above procedure is followed for unknown solution, in triplicates (table 2). Average absorbance is taken for the three values. For the absorbance value the corresponding concentration is read from the standard graph.

Uses of UV Spectroscopy

- The most ubiquitous use of UV is as a detection device for HPLC
- Quantification of metal ions in blood. E.g., calcium, iron, etc.
- Detection of exchangeable magnesium in soil.
- Quantitative estimation of composition of metals in alloys.
- Determination of copper in lubricating oils.
- Study of enzyme kinetics and elucidation of mechanisms.
- Estimation of hardness of water.
- Study of charge transfer complexes and coordination chemistry.

Limitations of UV Spectroscopy

- It can not be used for deriving structure of a molecule.
- Sensitive to impurities and solvent polarity.
- The compound must have UV-absorption.

IR Spectroscopy

Infrared spectroscopy deals with molecular transitions carried out by infrared region of the electromagnetic spectrum. It is useful to divide the infrared region into three sections; near, mid and far infra red. The most useful I.R. region lies between 4000 - 600cm^{-1}.

Region	Wavelength Range (μm)	Wavenumber Range (cm^{-1})
Near	0.78-2.5	12800-4000
Middle	2.5-50	4000-200
Far	50-1000	200-10

Photon energies associated with this part of the infrared (from 1 to 15 kcal/mole) are not large enough to excite electrons, but may induce vibrational excitation of covalently bonded atoms and groups.

Principle

Hooke's Law

Molecular vibration follows simple harmonic oscillation. According to Hooke's law, "for vibrating molecule force of vibration is directly proportional to displacement."

Molecular Vibrations

Equilibrium Bond Length   **Stretched**   **Compressed**

$$E_{vib} = \left(v + \frac{1}{2}\right)h\nu_o$$

v = vibrational quantum number. ν_o=fundamental vibrational frequency. h = Plank's constant.

When the molecule is in ground state (v=0) the energy of vibration is ½hν_o. It indicates even at absolute zero temperature molecules have vibrational energy and it is known as zero point energy. Thus molecules are not at rest even at absolute zero temperature. The fundamental vibrational frequency of a simple harmonic oscillator is given by

$$\nu_o = \frac{1}{2\pi}\sqrt{\frac{k}{\mu}}$$

$$\bar{\nu} = \frac{1}{2\pi c}\sqrt{\frac{k}{\mu}}$$

$\bar{\upsilon}$ = wavenumber , k = force constant, μ = reduced mass.

Virational spectra of homonuclear diatomic molecules are observed. It is because no dipole moment change during vibration.

Problem

The wave number of the fundamental vibration of Br79-Br81 is 323.2 cm-1. Calculate the force constant.

$$\mu = \frac{m_1 + m_2}{m_1 m_2}$$

$$\mu = \frac{78.9183 \text{ gmol}^{-1} \times 80.9163 \text{ g mol}^{-1}}{(78.9183 + 80.9163) \text{ g mol}^{-1} \times 6.023 \times 10^{23} \text{ mol}^{-1}}$$

$$= \frac{78.9183 \text{ gmol}^{-1} \times 80.9163 \text{ g mol}^{-1}}{159.8346 \times 6.023 \times 10^{23}} = 6.6333 \times 10^{23} g$$

$$= 6.6333 \times 10^{26} \text{ kg}$$

$$\bar{\upsilon} = \frac{1}{2\pi c} \sqrt{\frac{k}{\mu}}$$

$$k = 4\pi^2 c^2 \bar{\upsilon} \, \mu$$

$$k = 4 \, (3.14)^2 \, (3 \times 10^{10} \text{ cm s}^{-1})^2 \, (323.2 \text{ cm}^{-1})^2 \, (6.6333 \times 10^{-26} \text{ kg})$$

$$= 2.46.095 \text{ kg s}^{-1} = 246.095 \text{ Nm}^{-1}$$

Vibrational Spectra is a Band Spectra

Vibrational spectra give bands instead of discrete lines. This is because vibrational transition is accompanied by rotational transitions. Actual spectrum is a series of closed packed lines leading to unresolvable bands.

Instrumentation (Block Diagram)

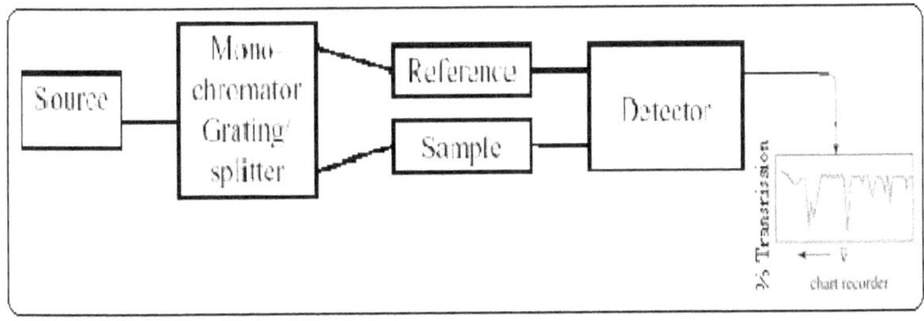

Radiation Source

- Nerst filament (ZrO and other rare earth oxides)
- Glober (Si-C)
- Ni-Cr wire
- Heated ceramic
- Mercury lamp

Sample Preparation

(i) Neat

Sample is used as for Recording IR spectra. The method is suitable for liquid and semi solid samples.

(ii) Nujol Mull

Nujol is a high boiling hydrocarbon. The sample is emulsified with nujol, spread in the sample cell and used for recording IR spectra.

(iii) Thin Film

In this method sample is dissolved in volatile solvents like chloroform. It is spread on the sample cell and allowed the solvent to evaporate. It leaves a thin film of sample which is used to record the IR spectra.

(iv) KBr Pellet

The sample is ground well in dry KBr crystals and pellet is made using pellet making dye. The pellet obtained is used for recording the IR spectra.

(v) Solution

The sample is dissolved in a suitable solvent and the solution as such is used for recording IR spectra.

Criteria for IR Activity

For a molecule to absorb IR, the vibrations or rotations within a molecule must cause a net change in the dipole moment of the molecule.

Homonuclear diatomic molecules like H_2, N_2, O_2, Cl_2, F_2, etc are not IR active. This is because no change in dipole moment in their vibrations.

Raman Spectra and IR Spectra

The criteria for a molecule to be IR active is polarisability change during vibration. So homonuclear diatomic molecules give Raman spectra. Thus Raman spectra and IR spectra are complementary to each other.

Molecular Vibrations

For linear polyatomic molecules the number of vibrations are 3N-5 and for non-linear molecules it is 3N-6. N=number of atoms.

The positions of atoms in a molecules are not fixed; they are subject to a number of different vibrations. Vibrations fall into the two main catagories of stretching and bending.

(i) Linear Triatomic Molecules (e. g., CO_2, N_2O, etc)

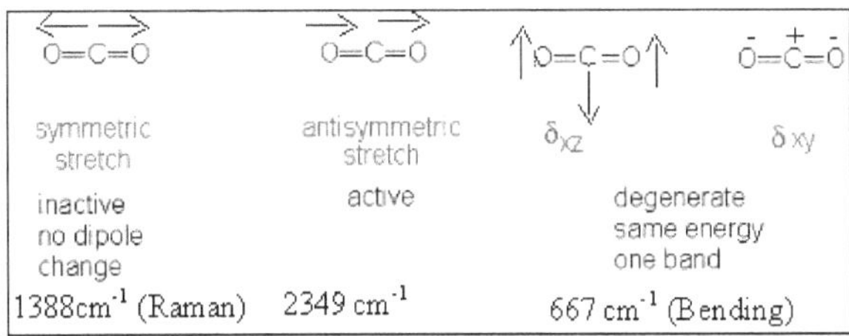

The number of vibrations is 4 (9-5).

(ii) Non-Linaer Triatomic Molecules (e.g., H_2O, SO_2, NO_2, etc)

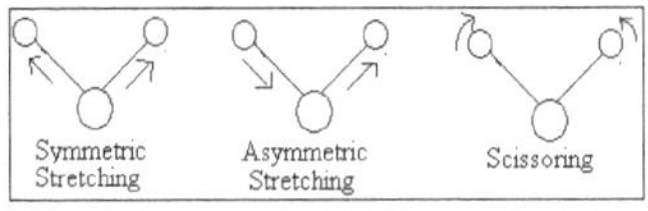

The number of vibrations is 3 (3N-6).

Vibrations of Methylene Groups

Methylene group in organic molecules have six modes of vibrations. Out of which two are stretching (symmetric and asymmetric) and four bending vibrations.

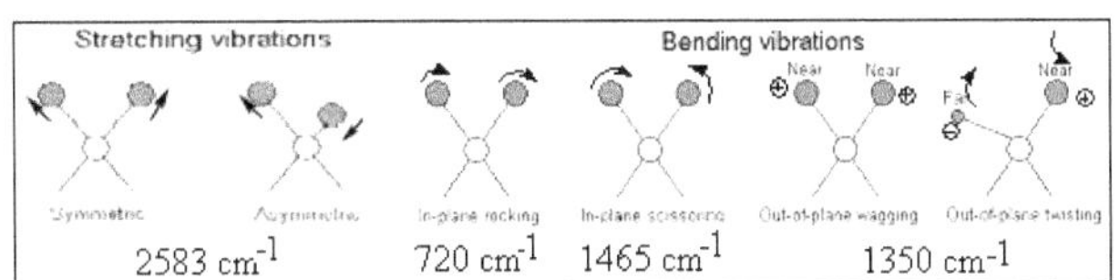

Selection Rule

IR transition to occur the change in vibrational energy level (Δv) should be ± 1.

Group Frequency Region

Absorption bands in the 4000 to 1450 cm^{-1} region are usually due to stretching vibrations of diatomic units, and this is sometimes called the group frequency region.

Finger Print Region

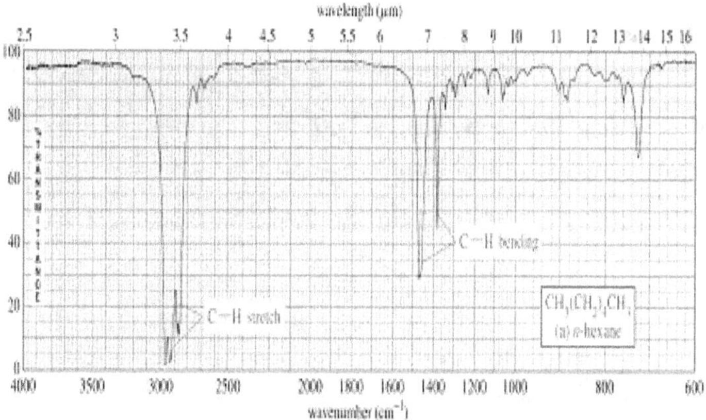

The complexity of infrared spectra in the 1450 to 600 cm^{-1} region makes it difficult to assign all the absorption bands, and because of the unique patterns found there, it is often called the fingerprint region. The region mostly corresponds to bending and rocking vibrations.

Fundamental Bands

IR absorptions corresponding to selection rule is called fundamental bands. In general most of the band on IR spectra is fundamental absorption bands.

Overtones

Absorption bands due to second and third vibrational level transitions are called overtones.

Combination bands

IR bands leading to the combination of two or more vibrations or difference of two vibrations are known as combination bands.

IR Absorption Regions for Different Functional Groups

Functional Group	Type	Vibration	Frequencies cm⁻¹
C-H	sp³ hybridized	$R_3C\text{-}H$	2850-3000
	sp² hybridized	$=CR\text{-}H$	3000-3250
	sp hybridized	C-H	3300
	aldehyde C-H	$H\text{-}(C=O)R$	2750, 2850
N-H	primary amine, amide	$RNH_2, RCON\text{-}H_2$	3300, 3340
	secondary amine, amide	$RNR\text{-}H, RCON\text{-}HR$	3300-3500
	tertiary amine, amide	$RN(R_3), RCONR_2$	none
O-H	alcohols, phenols	free O-H	3620-3580
		hydrogen bonded	3600-3650
	carboxylic acids	$R(C=O)O\text{-}H$	3500-2400
CN	nitriles	RCN	2280-2200
CC	acetylenes	R-CC-R	2260-2180
		R-CC-H	2160-2100
C=O	aldehydes	$R(C=O)H$	1740-1720
	ketones	$R(C=O)R$	1730-1710
	esters	$R(CO_2)R$	1750-1735
	anhydrides	$R(CO_2CO)R$	1820, 1750
	carboxylates	$R(CO_2)H$	1600, 1400
C=C	olefins	$R2C=CR_2$	1680-1640
		$R2C=CH_2$	1600-1675
		$R2C=C(OR)R$	1600-1630
-NO₂	nitro groups	RNO_2	1550, 1370

Some General Trends

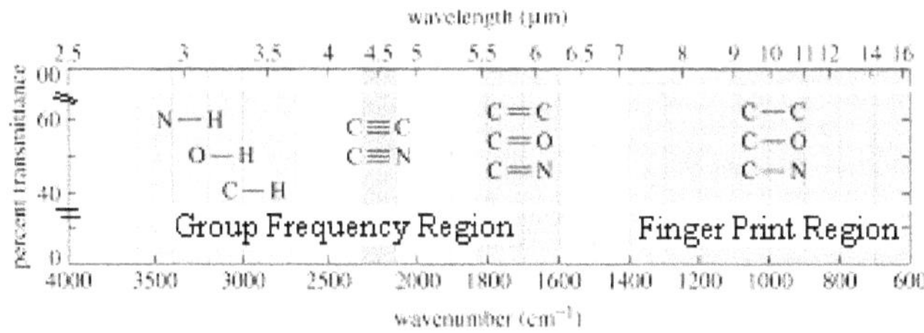

- Stretching frequencies are higher than corresponding bending frequencies. (It is easier to bend a bond than to stretch or compress it).

- Bonds to hydrogen have higher stretching frequencies than those to heavier atoms.

- Triple bonds have higher stretching frequencies than corresponding double bonds, which in turn have higher frequencies than single bonds.

- No two molecules will have exactly same IR spectrum, except enantiomers.

- Simple stretching frequencies appear between 3500 and 1600 cm^{-1}.

- Complex vibrations occur in the region 1400-600 cm^{-1}.

- Stretching frequency increases with increasing atomic masses of the atoms.

Bond	Bond Energy [kcal (kJ)]	Stretching Frequency (cm^{-1})
	Frequency dependence on atomic masses	
C—H ⎤ heavier	100 (420)	3000 ⎤
C—D ⎥ atoms	100 (420)	2100 ⎥ $\bar{\nu}$ decreases
C—C ⎦	83 (350)	1200 ⎦

(i) Increase in bond energy (strength) increase the IR stretching frequency.

Bond	Single (C-C)	Double (C=C)	Triple (C≡C)
Frequency (cm^{-1})	1200	1660	2200

Uses of IR Spectrum

- Identification of functional Groups.

- Absorption follows Bear's law. So can be used for quantitative analysis.

- Spectral matching can be done using library spectra and identify the compounds.

- Non-destructive analysis

Limitations of IR Spectra

- IR alone can not be used for structural elucidation.

- Functional group analysis based on IR spectrum is indicative but not conclusive.

- Molecular vibration should be IR active.

Colorimetry

Colorimetry is concerned with the visible region of electromagnetic spectrum (400-700 nm).The instrument used for measuring visible radiant energy is called colorimeter.

Principle

Coloured substances absorb radiations in accordance to Beer-Lambert's law.

From the colour comparison method concentration of unknown sample can be determined.

Methods in Colorimetry

There are four techniques used for the quantitative estimation of compounds by colorimetry.

1. Multiple Standard Method

A series of standard solutions are prepared. For example 0.10, 0.12, 0.14, 0.16, 0.18 and 0.20 g per litre. The colour of unknown sample is compared with the standard. Suppose the colour of the standard is in between 0.16 and 0.18 g/litre, then substandard (0.164, 0.168, 0.172, etc) is prepared and repeated the experiment to get more accurate concentration.

2. Duplication Method

Unknown sample is taken in a Nessler's tube and appropriate coloring reagent is added into it. In another Nessler's tube colouring reagent is taken little lower than the mark. To this standard, colouring reagent is added from a burette till the colour matches with the unknown. From the volume of standard added the unknown concentration is calculated.

3. Dilution Method

Standard and unknown sample are taken in two identical Nessler's tube. Light from the same source is allowed to pass through both the tubes. The intensity of the emerging light is compared. The more concentrated solution is diluted till both are having similar intensity of emission. From the volume of solution added for the dilution unknown concentration is calculated.

4. Balancing Method

In this method colorimeter is used. Standard and unknown solution are taken in a flat transparent tubes. A transparent plunger is kept in each tube and kept inside a calorimeter. By moving the plunger up and down the colour of both the tubes becomes identical. From the reading of the depth of the sample the unknown concentration is calculated.

Estimation of Iron by Colorimetry

A standard Fe^{2+} solution is prepared using ferrous ammonium sulphate. From this Fe^{2+} concentrations 5, 10, 20, 40, 60, 80, 100 ppm is prepared.

To 1 ml of the above standard solution add 1 ml each of 10% hydroxylamine and 0.2% sodium acetate solutions are added and mixed well. Further, 0.5 ml of 0.25% 1,10-phenanthralein solution is added well andallowed to stand for 5 minutes. A blank experiment is carried out using 1 ml of distilled water instead of sample.

Appropriate amount of distilled water is added in order to maintain constant volume in all the experiments. A similar procedure is followed for the unknown sample also. By using any of the visualization method of colorimetry get the accurate concentration of the unknown sample.

Flame Photometry

Flame photometers differ in three important ways from other instrumental techniques.

First, the power source and the sample-holder function are combined in the flame.

Second, in most applications of flame photometry, the objective is measurement of the sample's emission of light the sample.

A nebulizer converts the liquid sample into a fine aerosol. It is injected into the flame. The atoms in flame absorb energy at a characteristic wavelength.

Third, flame photometry is the only technique which can determine the concentrations of pure metals.

Principle

When elements are thermally excited it goes to higher energy state.

The distribution of atoms in the ground and excited state is governed by Boltzmann distribution law.

$$\frac{N^*}{N_0} = Ae^{-\Delta E/kT}$$

N^* = Number of atoms in the excited state.

N_0 = Number of atoms in the ground state.

A = Constant depends upon the element.

ΔE = Energy difference between ground and excited states.

K = Boltzman constant.

T = Temperature of the flame.

From the above equation it is clear that temperature of the flame controls the population of atoms in the excited state.

Instrumentation (Block Diagram)

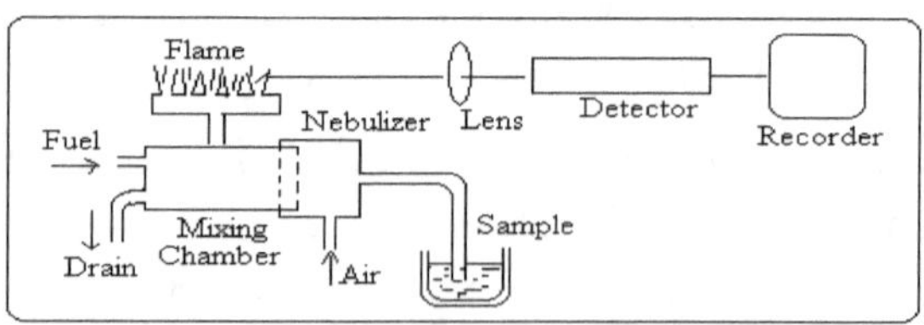

Estimation of Sodium by Flame Photometry

The principle of the experiment is that the sodium ions emit in the wavelength at 589 nm when exited in flame phtometer.

A standard solution of NaCl is prepared using deionized water. A series of dilutions are made from the stock solution of NaCl.

The flame photometer is switched on and the flamed it and adjusted to the blue flame.

Zero correction is carried out by aspirating deionized water through the aspirator and setting wavelength at 589 nm.

The standards are aspirated into the flame and the constant reading of emission is read from the display and tabulated the data. A standard graph is drawn between concentration and emission intensity. It gives a straight line passing through origin.

Similarly, sample and Unknown solution are also aspirated and taken the reading. The corresponding concentration is read out from the standard graph.

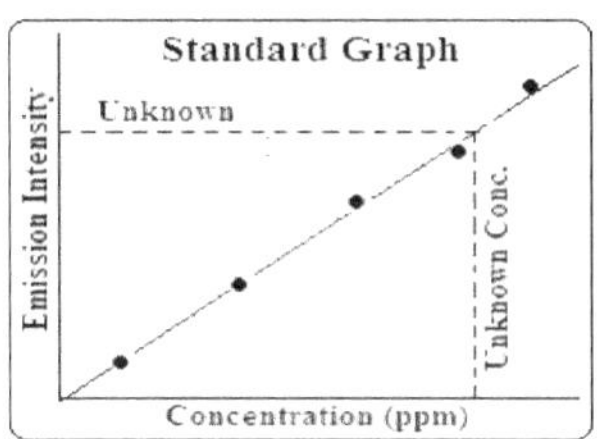

Uses of Flame Spectroscopy

- Alkali and alkali earth metals can be quantified by flame photometer.
- Analysis of metal ions in biological sample can be carried out quantitatively.
- Adulteration of fuels can be detected.
- Purity of metals and composition of alloys can be analyzed by flame photometry.
- Hardness of water can be quantified by flame photometry method.

Limitations of Flame Spectroscopy

- It is not suitable for all elements in the periodic table.
- The type of ionic species can not be determined.
- The sample must be solution in polar solvents preferably in water.
- Structural characteristics of the element can not be obtained.
- It is a destructive process and thus sample will be lost.

Atomic Absorption Spectroscopy

Atomic absorption spectroscopy in analytical chemistry is a technique for determining the concentration of a particular metal. Atomic Absorption Spectroscopy can be used to analyze the concentration of over 62 different metals in a solution.

The phenomenon of atomic absorption (AA) was first observed in 1802 with the discovery of the Fraunhofer lines in the sun's spectrum. It was not until 1953 that Australian physicist Sir Alan Walsh demonstrated that atomic absorption could be used as a quantitative analytical tool.

Origin of Atomic Spectroscopy

The metallic species is allowed to ionize in a flame. It absorbs light and goes to exited energy state. The amount of energy absorbed is directly proportional to the concentration (total number of atoms).

$$\text{The amount of light absorbed} = \frac{\pi e^2}{mc} \, nf$$

e = Charge of an electron
m = mass of an electron
c = speed of light
n = total number of atoms
f = oscillator strength

Atomic absorption methods measure the amount of energy (in the form of photons of light, and thus a change in the wavelength) absorbed by the sample. The electrons of the atoms in the flame can be promoted to higher orbital for an instant by absorbing a set quantity of energy (a quantum). This amount of energy is specific to a particular electron transition in a particular element. As the quantity of energy put into the flame is known, and the quantity remaining at the other side (at the detector) can be measured, it is possible to calculate how many of these transitions took place, and thus get a signal that is proportional to the concentration of the element being measured.

Block Diagram of Atomic Absorption Spectrometer

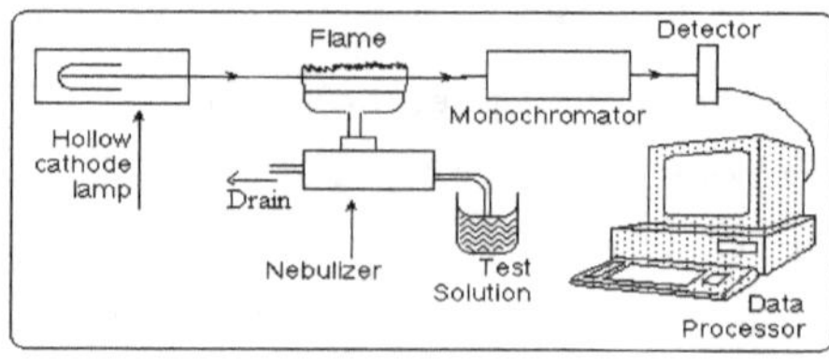

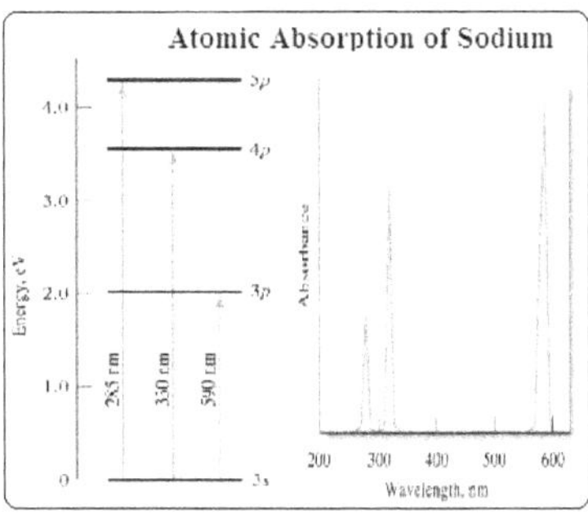

Types of Atomizer

There are two methods of adding thermal energy to a sample.

- Atomizers such as a graphite furnace. This technique typically makes use of a flame to atomize the sample.

- Plasmas, primarily inductively coupled plasmas, are also used to atomize the sample.

Light Source

The light source is usually a hollow-cathode lamp of the element that is being measured.

Atomic absorption spectroscopy can also be performed by lasers, primarily diode lasers. This technique is then either referred to as diode laser atomic absorption spectrometry (DLAAS or DLAS). It is also known as since wavelength modulation absorption spectrometry since wavelength modulation most often is employed in laser technique. Lasers are also used for multi element analysis.

Flames Used in Atomic Absorption Spectroscopy

Fuel	Oxidant	Flame Temperature (°C)
H_2	Air	2100
H_2	O_2	2800
Acetylene	Air	2200
Acetylene	O_2	3000

Hollow Cathode Tube

- The hollow cathode lamp (HCL) is a cathode made of the element of interest in the analysis, with a low internal pressure of an inert gas.

- A low electrical current ($\sim$ 10 mA) is applied to HCL. It allows the excitation of few metal atoms and emits a characteristic of that element.

- The light is emitted directed through the lamp's window and ultimately to the detector.

Nebulizer

- The nebulizer chamber thoroughly mixes acetylene (the fuel) and oxidant (air or oxygen).

- It creates a negative, plastic nebulizer tube which helps to suck liquid sample the nebulizer chamber, a process called aspiration.

- A small glass impact bead and/or a fixed impeller inside the chamber creates a heterogeneous mixture of gases (fuel + oxidant) and suspended aerosol (finely dispersed sample).

- This mixture flows immediately into the burner head where it burns as a smooth, laminar flame.

Liquid sample not flowing into the flame collected at the bottom of the nebulizer chamber a waste tube to a waste container. The waste collected will be highly acidic solution.

Estimation of Nickel by Atomic Absorption Spectroscopy

The principle is nickel sample on atomization in flame photometer absorbs radiant energy from source of light. It is concentration dependent and follows Beer-Lambert's law.

Nickel cathode lamp with filter for 232 nm is used as energy source.

Acetylene-oxygen flame is used for atomization of the sample.

$Ni(NO_3)_2$ can be used as a standard. A stock solution can be prepared by weighing known concentration and dissolving in minimum amount of dilute (1:1) nitric acid. Made up to required volume by 1% nitric acid solution.

A blank correction is done using deionized water.

A standard graph is drawn using the known concentration of nickel sample.

From the emission intensity of unknown sample concentration can be obtained from standard graph.

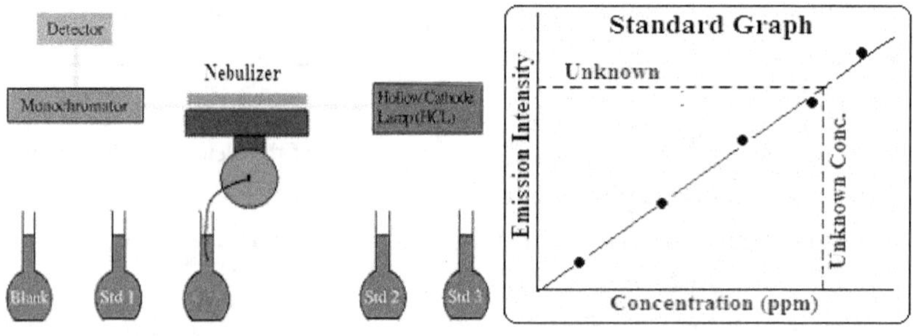

Uses of Atomic Absorption Spectroscopy

- Quantitative estimation of elements even at its native state.
- Simultaneous determination of multi-element mixtures.
- Analysis of biological fluids and the presence of trace elements.
- Analysis of metallic contamination in pharmaceutical preparations.
- Detection and quantification of elements in environmental and industrial emissions.
- Soil analysis and nutrient requirement planning.
- Detection of lead in fuels.

Limitations of Atomic Absorption Spectroscopy

- Elemental range is limited to metals and metalloids.
- Sample preparation is tedious and time consuming.
- The sample is destroyed by the analysis.
- Only one element at a time can be measured.

Assignment Questions

1. Calculate the energy associated with light having wavelengths 150 nm, 200 nm, 250 nm, 300 nm, 400 and 500 nm. Which is the wavelength most suitable to bring an excitation of energy gap 70 kcal/mol.
2. Carbonyl group absorbs at 1700 cm^{-1}. Calculate the force constant assuming the vibration is simple harmonic oscilation.
3. model question
4. Bring out the usefulness of wavelength of absorption and intensity of absorption in UV spectrum.
5. Comment on the use of atomic absorption analysis in environmental pollution monitoring.

Self Test

1. State and explain Beet Lambert's Law.
2. Explain the concept of interaction of electromagnetic radiations with molecules.
3. List out the electronic transitions possible to carbonyl group and double bonded compounds.
4. Bring out the use of Woodward-Fiesher rule with an example.
5. Analyze the principle of IR spectroscopy which suits to molecular vibrations.
6. Define group frequency, fundamental band and finger print region with examples.
7. Distinguish colorimetry from spectrophotometry.

8. Explain the terms Nebulizer, atomizer, monochromator.

9. Why metals have characteristic atomic absorption?

10. Determine the amount of Fe^{2+} present in a given sample using appropriate spectroscopic technique.

TWO MARK QUESTIONS AND ANSWERS

1. What is an electrochemical equation? Give an example.

    Electrochemical equation is a chemical equation which indicates the electron flow in the reaction. It is either oxidation or reduction reaction. Invariably electrons are involved in the reaction.

$$Zn \longrightarrow Zn^{2+} + e^-$$

2. Voltmeter cannot be used to measure the emf of a cell-why?

    Voltmeter consumes appreciable amount of current. So it is not suitable for the measurement of emf of a cell.

3. Among the following metal couple which metal displace the other metal from its salt solution?

    Pb-Fe; Zn-Mg

    Pb displaces iron. Zinc displaces magnesium. In electrochemical series Pb and Zn are having higher electrode potential than Fe and Mg.

4. Calculate the emf when SHE is coupled with SCE.

    SHE $E° = 0$

    EMF = Rediction potential of RHE-Reduction potential of LHE

    $0-(- 0.242) = 0.0242$ V.

5. Give the applications of ion selective electrodes.

    Determination of anions and cations. e. g., F^-, NO_3^-, Ca, K, etc.

    Pollution monitoring in industrial discharges and emissions..

    Body fluid analysis, like blood sugar quantification

    Analysis of combustion products.

6. What are the medical uses of membrane electrodes?

    Quantification of biological compounds in blood, urine, etc.

    Monitoring of enzyme reactions.

7. Write the electrode reaction of Ag-AgCl electrode.

$$Ag^+ + e^- \longrightarrow Ag$$
$$Cl^- \longrightarrow Cl + e^-$$

8. Write the Nernst equation useful for the measurement of pH.

$$E = E° + RT/nF \ pH.$$

9. List the advantage and disadvantages of potentiometric titrations.

Advantages

- Coloured solutions can be used in the estimation.
- No indicator is required.

Disadvantages

- There should be an electrochemical reaction.
- Selection of appropriate electrode.

10. On titration of titration of mixture of strong and weak acids with base strong acid gets neutralized first. Why?

Weak acids are weakly ionized. Strong acids are completely ionized. So strong acid gets neutralized first then the weaker acid is neutralized.

11. Give the chemical basis of corrosion.

Metals tend to be more stable in its compound form than its native form.

Spontaneous reaction (ΔG = -ve) occurs which lead to corrosion.

12. What is rust and give its chemical constituents?

Rust is a hydrated ferric oxide. It is mixed oxides of iron. It is either hydrated ferric oxide (Fe_2O_3-$Fe(OH)_3$ or mixture of ferrous and ferric oxide ($FeO.Fe_2O_3$).

13. Corrosion is reverse of metallurgy-explain.

Metallurgy is the process of manufacturing pure metal from its ore (compound).

Corrosion is the formation of compound from pure metal.

14. What is a liquid-metal corrosion? Give an example.

Liquid on contact with solid metal surface dissolution occurs

e.g., dissolution of metals in Hg.

Dissolution of Pb in Ag at high temperature.

15. Why anode gets corroded rather than cathode?

Anode is the source of electron. In the process it forms ions. The ions get undergo reaction leading to corrosion. Cathode accepts electrons.

16. What is the role of oxygen in corrosion?

In the absence of moisture oxygen carries out oxidation corrosion.

$$M + O_2 \longrightarrow M^+ + O_2^-$$

17. Among the following selected the best design in terms of corrosion protection.

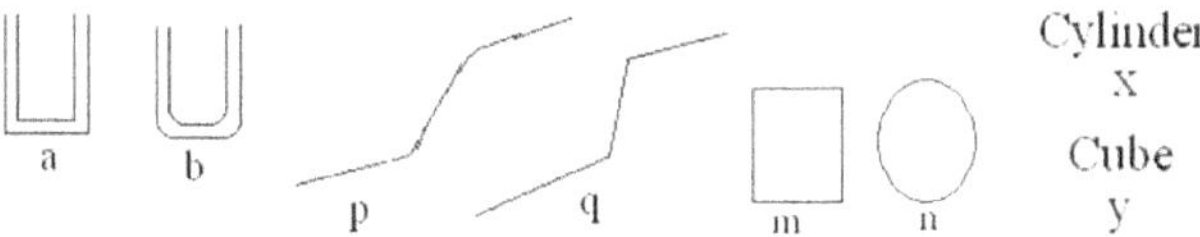

b, p, n and x. Less sharp edges. So corrosion will be less.

18. Among Cu and Au which is the preferred metal to be coated on Fe.

Au is preferred because in Galvanic series it is less active (noble) than Cu.

19. In gold plating alloy of gold with cobalt is used-why?

- Increase wear and tear resistance
- Decrease the cost.

20. What is the function of reducing agent in electroless plating?

Reducing agent spontaneously liberates metal from its compound. In the process platting occurs on the surface of the article. It initiates and participate in the internal redox reaction. So external current supply is not required in electrless plating.

21. Compare electroplating and electroless plating.

Sl.No.	Elelctroplating	Electroless Plating
1	An external energy source source initiate electrochemical reaction.	Internal redox reaction performs the plating process.
2	Separate anode and cathode	Catalyst acts as anode.
3	Quality it less compare to electroless plating.	High quality.
4	Coating materials to be electrically conductive.	Need not be the case.
5	More suitable for metal coatings.	Even non-metals can be coated.

22. Is there any relation between calorific value and molecular structure? Justify the answer.

$$HCV = \frac{1}{100}\left[8080 \times C + 34500 \left(H - \frac{O}{8} \right) + 2240 \ S \right] \ kcal/kg$$

In the above equation there is no relation with respect to structure. In actual case there is a relation between calorific value and molecular structure.

Aromatic compounds will have less calorific value. It is the basis for cetane number.

23. Low calorific value is more significant than high calorific value-why?

It is more realistic in practical use internal combustion engine where the products of combustion is allowed to escape. The heat consumed as latentent heat of vaporization is taken in to account in LCV.

24. Among C, H and S which one contributes more to the calorific value.

$$HCV = \frac{1}{100}\left[8080 \times C + 34500 \left(H - \frac{O}{8} \right) + 2240\, S \right] \text{kcal/kg}$$

Carbon contributes highest for calorific value. It is because percentage of carbon in is higher than other elements.

25. Write the principle of sulphur estimation in ultimate analysis.

Sulfur is oxidized to sulphate and is precipitated using $BaCl_2$.

$$S \text{ (in coal)} + O_2 \longrightarrow SO_4^{2-}$$
$$SO_4^{2-} + BaCl_2 \longrightarrow BaSO_4 \downarrow$$

26. What is heavy oil? Give its uses.

- It is the oil coming out towards the end of fractional distillation of petroleum.
- It contains high boiling aromatic and polycylic compounds.
- It is used in catalytic cracking.

27. Write the function of catalyst in catalytic cracking.

The catalyst used in catalytic cracking process brings down cracking temperature.

It initiate and sustain free radical reaction.

28. TED is not advisable to use as an antiknocking agent-why?

Lead is an environmental pollutant.

Decreases the engine life.

29. A fuel is rated as 75 % octane equivalent. Deduce its meaning.

The knocking behavior of the fuel is equivalent to 75 % isooctane and 25 % n-octane.

30. What is the principle behind Orsat flue gas analysis?

Known volume of gaseous emissions is allowed to react with KOH (for CO_2), alcoholic pyrogallol (for O_2) and Ammoniacal Cu_2Cl_2 (for CO). From the decrease in volume of the gas, the percent contributions of individual components are calculated.

31. Among C, H and S which consume highest amount of O_2 for its complete oxidation.

Hydrogen consumes highest amount of O_2.

32. What are the disadvantages of a knocking fuel?

- Noise pollution.
- Low efficiency of the fuel
- Decrease engine life.
- High emissions.

33. Define degrees of freedom and give its significance.

It is the minimum number of independent variables such as temperature, pressure and concentration of a phase to be specified in order to define a system completely. It gives the equilibrium temperature and pressure or composition.

34. Is it possible to use phase rule for a homogenous equilibrium? Why?

Phase rule is applicable only to heterogeneous equilibrium.

In homogeneous equilibrium it is not possible to identify the phases.

35. What is the need for condensed phase rule?

It is a simplification of phase rule. It allows to use the open system experiments. So vapour need not be considered.The phase diagram also reduced from three dimensional (composition, temperature and pressure) to two dimensional one (temperature and composition).

36. What are the phases in equilibrium at eutectic point?

Liquid metals and eutectic mixture.

37. Describe the characteristics of bivarient equilibrium in Ag-Pb system.

It is the liquid phase. Both Ag and Pb are in homogeneous liquid and one phase.

$$F = C - P + 1$$
$$F = 2 - 1 + 1 = 2.$$

38. Comparative analysis of normalizing and hardening.

Sl. No.	Normalizing	Hardening
1	Any steel quality can be used	Steel with high carbon content can not be used.
2	Depending upon carbon content heating temperature is decided.	Heated above 1000°C.
3	After heating cooled in air.	After heating cooled adiabatically (Suddenly)
4	Improves fine structure of steel.	Improves abrasion resistance.

39. What is the role of carbon in steel?

Carbon provides different physicochemical properties to iron depending upon the percentage. During heat treatment different type of metalloids are formed which are having specific properties for specific applications.

40. Write the composition and use of constantan.

It is a Cu-Ni alloy. Cu = 60 % and Ni = 40 %.

It is used in Resistance wires and thermocouples.

41. How many bivarient equilibrium possible in water system? What are they?

There are three bivarient systems possible in water system. They are Solid-ice, liquid-water and water-vapour. All are pure water systems.

42. What happens if a liquid having 60% silver and 40 % lead on cooling?

On cooling silver crystallized out first. On further cooling at 303°C Eutectic mixture is formed and system becomes completely solid.

43. Distinguish absorption and emission spectroscopy.

Sl. No.	Absorption Spectroscopy	Emission Spectroscopy
1	Spectrum is obtained due to absorption of incident radiation.	Emitted radiation is recorded as spectra.
2	Beer Lambert's law applies.	It is not applicable.
3	Absorption spectra is recorded 180° to the incident light.	Emission spectra is recorded 90° right angle to the incident light.
4	e.g., UV-Spectra, IR spectra.	Atomic absorption spectra, Fluorescence spectra.

44. Below 250 nm glass cuvette cannot be used to record UV spectrum. Why?

Glass absorbs strongly below 250 nm. So quartz cuvette has to be used for recording VU spectra below 250 nm.

45. Two solutions of a same compound gave absorbance 0.103 and 0.053 against a same blank. Is their concentration is related to 1:0.5. Justify your answer.

Yes. It is because of Beer-Lambert's law there is linear relationship between concentration and absorbance.

46. Give the order of relative energies of electronic transition of different bonded electrons in a compound.

$$\sigma \rightarrow \sigma^* > \sigma \rightarrow \pi^* > \pi \rightarrow \pi^* > n \rightarrow \sigma^* > n \rightarrow \pi^*$$

47. β-Carotene is yellow in colour. Why?

It has 11 conjugated double bonds. Conjugation increases the wavelength of absorption. So it absorbs in visible region and hence coloured.

48. What is the significance of reduced mass in IR frequency?

$$\mu = \frac{m_1 + m_2}{m_1 m_2} \qquad \upsilon_o = \frac{1}{2\pi}\sqrt{\frac{k}{\mu}}$$

Reduced mass and frequency of absorption are inversely related.

49. Calculate UV maximum for the following compound.

Base value	= 210 nm
4 alkly groups	= 20 nm
Exocyclic Double bond	= 5 nm
Total	= 235 nm

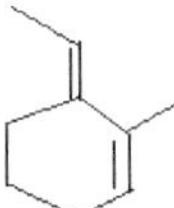

50. Draw the block diagram for flame photometer.

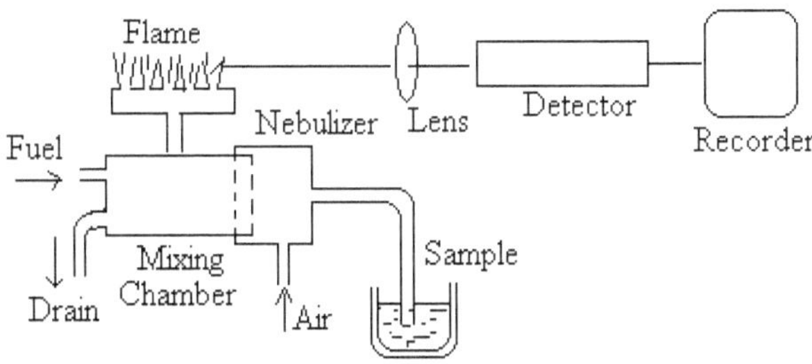

51. What are the flames used in atomic absorption spectroscopy and what is its function?

Fuel	Oxidant	Flame Temperature (°C)
H_2	Air	2100
H_2	O_2	2800
Acetylene	Air	2200
Acetylene	O_2	3000

The energy of the flame is used to atomize the sample atoms and excite it to higher energy level.

52. Distinguish zero correction and baseline correction.

Sl. No.	Zero Correction	Baseline Correction
1	It is done for a single wavelength.	It is performed for the entire wavelength of recording a spectrum.
2	It is used when doing quantitative estimation.	It is used when recording spectra of a sample.
3	Baseline correction can not be performed in simple single beam instruments.	Double beam instruments can perform baseline correction.
4	It is the background correction against black for a particular wavelength.	It is the background correction against blank for entire wavelength of recording a spectrum .

53. What are the applications and limitations of flame spectroscopy?

Uses: (i) Quantitative estimation of elements even at its native state.

(ii) Similtaneous determination of multi-element mixtures.

Limitations: (a) Elemental range is limited to metals and metalloids.

(b) Sample preparation is tedious and time consuming.

SOLVED PROBLEMS

1. Calculate the standard free energy change for the following reaction?

$$Fe^{2+} + Ag^+ \rightarrow Fe^{3+} + Ag \qquad [E^°_{Fe^{3+}/Fe^{2+}} = 0.77V \text{ and } E^°_{Ag^+/Ag} = 0.80 \text{ V}$$

$\Delta G^° = -nFE^°$

$E^° = E^°_{right} - E^°_{left}$

$\quad = 0.80 \text{ V} - 0.77 \text{ V} = 0.03 \text{ V}$

$\quad = 1 \text{ mol} \times 96500 \text{ C mol}^{-1} \times 0.03 \text{ V}$

$-2895 \text{ J} = 2.895 \text{ kg}$

2. Write a Nernst equation to calculate eht reduction potential of O2 at pH = 7. The partial pressure o O2 = 0.20 bar and E° = 1.229 V at pH = 7.

$$2H^+ + \frac{1}{2}O_2 + 2e^- \rightleftharpoons H_2O$$

$$E = E^° - \frac{0.0592}{2} \log \frac{1}{[H^+][P_{O_2}]^{\frac{1}{2}}}$$

$$E = 1.229 \text{ V} - 0.0296 \text{ V} \log \frac{1}{(10^{-7})^2 (0.2)^{\frac{1}{2}}}$$

$\quad = 1.229 \text{ V} - 0.0296 \text{ V} [14.0 - 0.4472]$

$\quad = 1.229 \text{ V} - 0.4011 \text{ V} = 0.828 \text{ V}$

3. Calculate the free energy change for the following cell. Comment on the result.

$\quad Fe^{2+}/Fe^{3+}//Ag^+/Ag$; $E^° Fe^{3+}/Fe^{2+} = 0.77$ V, $E^°Ag^+/Ag = 0.8$ V.

$E^° = E^°_{right} - E^°_{left}$

$0.80 - 0.77 = 0.03$ V.

$\Delta G = -nFE^°$

$\quad = -1 \times 96500 \text{ C mol-1} \times 0.03 \text{ V}$

$\Delta G = -2895 \text{ J} = -2.895 \text{ kJ}$.

$\quad$ Negative sign of ΔG value indicates the reaction is spontaneous.

4. The emf of a glass electrode with SCE at 298 K in a given solution is 0.4188 V. Calculate the pH of the solution.

$\quad E = E^°-0.0592 \text{ pH}$

$\quad 0.4188 = 0.248 - 0.0592 \times \text{pH}$

$\quad 0.4188 = 0.1888 \times \text{pH}$

$$pH = \frac{0.4188}{0.1888} = 2.22$$

5. IR spectrum of a compound showed finger print absorption bands in the following frequency region. Assign the values to the functional groups in the molecule.

3300 cm-1 (broad) , 2920 cm-1 2770 cm-1 (sharp) and 1720 cm-1 (strong).

3000 cm-1 = OH group or NH2 group.

2920 cm-1 = CH stretching

2770 cm-1 = Aldehyde CH stretching

1720 cm-1 = Carbonyl stretching.

QUESTION BANK

PART A QUESTIONS

1. Define oxidation half cell and reduction half cell.
2. What is a cell reaction? Give an example.
3. Distinguish Galvanic cell and electrolytic cell.
4. Differentiate cell and battery.
5. What is a salt bridge and give its functions?
6. Define reversibility and irreversibility in terms of electrochemistry.
7. State Pogendorff's compensation principle and give its applications.
8. Define electrode potential and compare it with standard electrode potential.
9. Define oxidation and reduction potential.
10. What is an electrochemical series and how is it constructed?
11. Construct a cell from the following data

    $Ag^+/Ag = 0.80$ V; $Fe^{2+}/Fe = -0.44$ V.
12. State the uses and limitations of electrochemical series.
13. Predict which metal liberate hydrogen from dilute acid.

    $Ca^{2+}/Ca = -2.87$ V; $M^{2+}/Mn = -1.03$ V; $Ag/Ag^+ = -0.80$ V; $Cu^{2+}/Cu = 0.34$ V
14. Using the following data write electrode reactions and cell reaction.

    $Mg^{2+}/Mg = -2.34$ V; $Cu^{2+}/Cu = 0,034$ V.
15. Give a Nernst equation to calculate equilibrium constant and explain the terms involved in it.
16. List out the conditions to be maintained in a standard hydrogen electrode.
17. What is the use of calomel in calomel electrode?
18. What are primary standard electrodes and secondary standard electrodes?
19. Give the principle of construction of glass electrode.
20. What are the specificities of ion selective electrodes?
21. What are membrane electrodes and the function of membranes?
22. What is the principle involved in potentiometric titrations?
23. Draw the graph of potentiometric titration of silver nitrate and potassium chloride and explain its characteristics.
24. Define conductivity and give its applications.
25. Bring out the principle behind conductometric titration.
26. Explain why weak acids and week bases can not be titrated using conductometry.

27. Compare the concepts around potentiometry and conductrometry.

28. Define corrosion with a suitable example.

29. What is the significance of corrosion with respect to environment and economy?

30. State Pilling-Bedworth rule.

31. What is hydrogen embrittlement and its implications?

32. How does dry corrosion and wet corrosion differ?

33. Classify the following as protective and non-protective oxides from the data.

34. Bring out the used in construction of Galvanic electrochemical series and the need for such series.

35. Based on Galvanic series predict the anode and cathode from the following metal pair.

      Mg-Fe; Cd-Ag.

36. What is the influence of pH and pollution on corrosion?

37. What is the concept behind decarburization and its importance?

38. Define pitting corrosion and its relevance to mechanical strength.

39. What is the principle behind differential aeration corrosion?

40. Bring out the criteria for Galvanic corrosion.

41. List out different corrosion control methods.

42. What is the concept behind sacrificial anode?

43. Define impressed current and its implications.

44. What is cathodic and anodic protection?

45. Enumerate the need for coating.

46. Define paints and its functions.

47. State the importance of pigments with examples.

48. What is the need for metallic coating and properties needed for the coating metal?

49. Bring out the principle and application of electrophoresis.

50. What are the salient features of electroless coating?

51. Define ion implantation and its uses.

52. How is metal spraying technology helps in pretension of corrosion?

53. Distinguish electroplating and electroless plating.

54. Define fuel and the characteristics of an ideal fuel.

55. What is calorific value and the need for the same?

56. Compare the calorific value of solid, liquid and gaseous fuels.

57. State the units of calorific value and give their interrelationships.

58. Classify fuels based on occurrence with examples.

59. List out the proximate analysis parameters.

60. What is the significance of ash in a fuel? Whether it varies in proximate and ultimate analysis?

61. 10 g of coal heated at 950°C for 7 minutes. On cooling it weighed 7.83 g. Calculate the percentage volatile matter in the coal.

62. How does the percentage of nitrogen in a fuel is determined?

63. Perform a need based distinction between proximate and ultimate analyses.

64. State the principle behind the manufacture of coke.

65. What is fractional distillation? List out the fractions collected in the fractional distillation of petroleum.

66. What is the need for vacuum distillation of petroleum and the fractions obtained in such process?

67. List out different cracking methods. Give its relative merits.

68. What is the role of catalyst in a cracking process?

69. Give the advantages of fluid bed catalytic cracking.

70. What are knocking characteristics and give its relevance to fuel quality?

71. What are anti-knocking agents and give its function?

72. Analyze the significance of rating a fuel by octane number.

73. What is the meaning of cetane rating?

74. Why there is a need for synthetic petrol?

75. What is Orsat analysis and give its uses?

76. Compare Bergius and Fisher-Tropsch processes.

77. Give an equation to calculate the theoretical air for a fuel.

78. Why nitrogen does not consume oxygen in burning process?

79. What is the principle behind the Orsat analysis of CO, CO_2 and O_2?

80. What is the chemical basis on which phase rule is formulated?

81. Calculate the maximum number of phases possible for a two component system.

82. Define invariant point and eutectic point.

83. Give the significance of isopeth and isobar.

84. What is the meaning of solidous cure and melting point curve?

85. What are the limitations of phase rule? Is it possible to over come it?

86. State liver rule and give its applications.

87. How many two phase equilibrium possible in water system?

88. What is the significance of melting point curve of water?

89. What is thermal analysis and give a typical thermal analysis graph?

90. Bring out the relationship between thermal analysis and phase diagram.

91. What is the meaning of simple eutectic system?

92. Explain a condensed phase system with an example.

93. What are the characteristics of invariant point in Pb-Ag system?

94. Analyze the principle involve in Pattinson's process of manufacture of pure silver.

95. List out the significance of alloys.

96. Distinguish impurity from alloying element.

97. What is the effect of alloying element on melting point and boiling point of the base metal?

98. Distinguish ferrous alloys from non-ferrous alloys.

99. Give the composition and uses of nichrome.

100. Define heat treatment. Give its relevance with respect to steel.

101. What are hypereutectoid and hypoeutectoid steels?

102. What are brass and bronze?

103. What are the characteristic features achieved by subjecting heat treatment on steel?

104. What is the carbon content in carburized steel?

105. What is an analytical technique and why it is explored in modern chemistry?

106. Classify the following parameters deduced by analytical and non-analytical techniques.

    (i) 10 g coal (b) boiling point (c) 60 % methane in coal gas (d) 98 % pure.

107. Define spectroscopy. Is it a physical technique or chemical technique or biological technique?

108. Bring out the meaning of absorption and emission in relation to spectroscopy.

109. Define electromagnetic radiation and give its energy relation.

110. State Beer-Lambert's law and give its applications.

111. Define absorption and extinction coefficient.

112. Give the UV visible range and the source of radiation.

113. What are single and double beam spectrophotometers?

114. List out the detectors used in spectroscopic techniques.

115. What is the function of a monochromator?

116. What is the consequence of Frank-Condon principle?

117. UV spectra are band spectra-why?

118. What is the contribution of chromophore in VU-visible spectroscopy?

119. What are inferences can be obtained from an UV-visible spectrum?

120.   From the following data determine theconcentration of the unknown solution.

Conc. (mg/l)	0	5	10	15	20	25	unknown
Absorbance	0.001	0.025	0.052	0.074	0.101	0.121	0.040

121.   State Hooke's law and give its applications.

122.   IR spectroscopy is vibrational and rotational transition in a molecule-Justify.

123.   What is the significance of force constant in IR spectrum?

124.   What is the radiation source in IR spectroscopy?

125.   What are IR active and IR inactive vibration modes?

126.   What are hypereutectoid and hypoeutectoid steels?

127.   What are brass and bronze?

128.   What are the characteristic features achieved by subjecting heat treatment on steel?

129.   What is the carbon content in carburized steel?

130.   Define electromagnetic radiation and give its energy relation.

131.   State Beer-Lambert's law and give its applications.

132.   Define absorption and extinction coefficient.

133.   Give the UV visible range and the source of radiation.

134.   What are single and double beam spectrophotometers?

135.   List out the detectors used in spectroscopic techniques.

136.   What is the function of a monochromator?

137.   What is the consequence of Frank-Condon principle?

138.   UV spectra are band spectra-why?

139.   What is the contribution of chromophore in VU-visible spectroscopy?

140.   What are inferences can be obtained from an UV-visible spectrum?

141.   What is the relation between molecular vibration and IR activity?

142.   What is the use of finger print region in IR spectrum?

143.   In IR spectrum carbonyl group absorption is strong-why?

144.   State the principle involved in colorimetric estimations.

145.   Define standard graph and give its uses.

146.   What is the need for doing blank experiments in analytical estimations?

147.   What are nebulizers and atomizers?

148.   Compare flame photometry and atomic absorption spectrometry.

149.   What is a standard solution and why is it required in quantitative estimation?

150.   What is the origin of signal in atomic absorption photometry?

151.   Flame spectroscopy is a emission spectroscopy-justify.

PART B QUESTIONS

1. Construct an electrochemical cell and explain the electron transfer reactions.

2. Depict the principle and the method of measurement of emf.

3. What are the salient features and limitations of ECS?

4. Explain the applications of ECS.

5. Construct a glass electrode and explain its use in measurement of pH.

6. Explain the construction and electrode reactions of SE.

7. Explain the use of SCE for the measurement of emf of an unknown cell.

8. Construct an ion selective electrode and show its use measurement of ionic concentration.

9. Outline the quantitative estimation of F3+ by potentiometry.

10. Perform an acid base titration by conductomerty.

11. Explain the electrochemical theory of corrosion.

12. Analyze dry corrosion mechanisms.

13. Explain the principle of construction of Gavanic series and its uses.

14. Explain the different type of corrosion pretension methods.

15. Elaborate the environmental factors influence the corrosion.

16. Enumerate the physicochemical parameters influence the corrosion.

17. Explain the different techniques involved in cathodic protection.

18. Write note on corrosion inhibitors.

19. Analyze the relative merits of different coating methods.

20. Describe the use of paints in the prevention of corrosion.

21. Explain the experimental details of electroplating of gold on steel.

22. Outline the electroless coating of nickel on silicon.

23. Describe the proximate analysis parameters for proximate analysis of coal.

24. Explain the principle of ultimate analysis of C, H, S and N.

25. List out the fractional distillation fractions of petroleum and its chief constituents.

26. Schematically explain the process of fluid bed catalytic cracking and list out its superiority over fixed bed catalytic cracking.

27. Explain the process of fixed bed catalytic cracking.

28. Describe cetane number method of fuel grading.

29. Explain the method of flue gas analysis.

30. What is the composition and use of CNG, producer gas and water gas?

31. Bring out the salient feature of manufacture of synthetic petrol by Fischer-Tropsh method.

32. Schematically describe the Bergius process of manufacture of gasoline.

33. Derive phase rule and give its applications.

34. Draw the phase diagram of water and explain its salient features.

35. Construct a phase diagram using a thermal analysis data.

36. Explain the simple eutectic phase diagram of Pd-Au system.

37. Explain the use and composition of non-ferrous alloys.

38. What are ferrous alloys? Give its properties with examples.

39. What are the different types of annealing of steel? Give its significance.

40. Subject steel to carburizing, nitriding and cyniding and describe the changes in the properties of steel in the above processes.

41. State and explain Beer Lambert's law and give its uses.

42. What are the components of double beam UV spectrometer and explain its functions.

43. Explain Frank-Condon Principle with its significance.

44. Graphically show the different types of electronic transitions in a molecule.

45. What are the parameters to be analyzed in an UV spectrum and their relevance to molecular structure?

46. Write note on Woodward-Fischer rules.

47. Perform a quantitative determination of Fe^{2+} spectrophotometrically.

48. Describe the relationship between molecular vibration and IR spectrum?

49. Explain the different methods of sample preparation for recording IR spectrum.

50. Analyze the number and modes of vibrations which are IR active in a molecule.

51. List out the important functional groups and its fundamental IR absorption bands.

52. Describe a suitable method of estimations of Fe^{2+} by colorimetric method.

53. Explain the experimental protocol involved in the estimation of sodium ions by flame photometry.

54. Analyze the use of atomic absorption spectroscopy for quantitative estimation with the help of an example.

PROBLEMS-BANK

1. Calculate the gross and net calorific value of a coal having the following composition.

   C= 80 %, H = 7 %, O = 4 %, S = 2.5 A%, N = 2.1 % and Ash = 4.4 %.

2. Calculate the volume of air required for the complete combustion of one litre of CO.

3. A producer gas the 30 % CO, 12 % H2, 4 % CO2, 2 % CH4 and 52 % N2 by volume. What is the composition of dry flue gas if 50 % excess air is used for burning 100 m3 of the gas?

4. A coal has a following composition by weight. C = 90 %, O =3 %, S = 0. 5 % N = 0. 5 % and ash = 2.5 %. Net calorific value of the coal is 8490 kcal/kg. Calculate the percentage of hydrogen and HCV?

5. An alloy AB of 10 g weight contained A at 25 %. The molten AB on cooling gave out B and the eutectic alloy with A and B equal percentage. What is the amount of B formed?

6. 1000 kg of argentiferrous lead containing 0.1 % silver is melted and then allowed to cool. If the eutectic contains 2.6 % Ag, what is the mass of eutectic formed and the mass of lead separated out?

7. Construct a phase diagram for an AB system from the following data.

Mass %B	0	10	25	35	45	55	65	80	92	100
Melting Point °C	119	112	99	90	78	83	95	108	113	115

8. Determine eutectic temperature and eutectic composition. Draw a cooling cure for the melt with composition 40 % B. and discuss.

9. A radio station broad cost its programs at 96.4 kHz. Calculate the wavelength emitted by the transmitter and the energy of radiation.

10. Calculate the emf of a Daniel cell at 25°C when the concentration of $ZnSO_4$ and $CuSO_4$ at 0.001 M and o,1 M respectively. The standard emf of the cell is 1.2 V.

11. Calculate the value of Faraday's constant if ΔG for Daniel cell is 212.3 kJ and emf is 1.1 V.

12. Write the electrode and cell reaction for the following.

   $Fe/Fe^{2+}//Fe^{2+}\text{-}Fe^{3+}/Pb$

   $Zn/ZnO_2^{2-},OH//HgO/Hg$

13. Calculate the reduced mass of CN group. (atomic mass of C = 12.011 and N = 14.0067).

14. Calculate G° and E° for the following using the given values.

   $$Cr^{3+} + 3e^- \rightarrow Cr \qquad E^0 = 0.50 \text{ V}$$

   $$Cr^{3+} + e^- \rightarrow Cr^{2+} \qquad E^0 = -0.41 \text{ V}$$

15. A 20 ml of mixture of strong acid and weak acid is subjected to neutralization using 0.05 M, NaOH solution. The process was monitored conduct metrically. The volume of base consumed for neutralizations are 15.0 ml and 4.5 ml. Calculate the amount of acids present in 1 litre of the acid mixture.

16. In a potentiometric titration 20 ml of 0.005 M Fe^{2+} solution is titrated with 0.005 M dichromate solution. A graph of $\Delta E/\Delta V$ versus ΔV shows a peak at 19.4 ml. Calculate the amount of Fe^{2+} in 1 litre of the given solution.

GUIDELINES TO STUDY

During the Semester Days

1. Read the subject every day immediately after it is discussed in the class.
2. During weak ends refresh the same.
3. Not to accumulate for the study leave.
4. In addition to class notes study one text book which available locally.
5. Further read relevant books from the library.
6. If any doubt or difficulty in understanding the concept clarify it as early as possible from the course instructor.
7. Attend all class tests, internal and model examinations.
8. Submit all assignments, class works, etc.
9. Complete all record and other writing works on time.
10. Regularly attend the class and participate in learning.

During Study Holidays (Before Examination)

1. Select most favorable unit in which there is a better understanding in comparison to other units first and thoroughly study it. Then go to next favorable unit. The most difficult one has to be read at the end.
2. From every unit it is fixed to have 20 marks questions. So concentrate and study the easy parts first.
3. In an unit partly known and partly find it difficulty, then make sure read well the known part instead of wasting time on unknown part in the last moment. In general question is to be uniformly distributed in the syllabus.
4. Keep the concepts in each title clear to avoid confusion and write wrong answer to the question.
5. Not to over burden on one day and feel uncomfortable in other day examinations.

EXAMINATION TIPS

PART A

1. Invariably answer all the questions without fail irrespective of answer is known or unknown.
2. If answer is known,
   a. Answer as brief as possible, not more than three sentences.
   b. Give simple explanation if asked for.
   c. Give examples if available.
3. If answer is unknown,
   a. Give a relevant answer if any such idea.
   b. If no idea at all about the answer write any thing about the subject mater but shall not be lengthy (about 3 to 4 lines).
   c. If absolutely no idea at all about the answer write any thing but shall not be lengthy (about 3 to 4 lines) even repeat the question once.
   d. What is written should be neat and clear irrespective of correctness or relevance to the question.
4. What is written should be neat and clear irrespective of correctness or relevance to the question.
5. Write the correct question number outside the margin.
6. Never leave a Part A question unattempted.

PART B

1. Start answering from the best known question first.
2. Before start answering the question make sure that answer to all the subsections of the question is known.
3. If one (part) of subsection answer is known in section (a) and section (b)
   a. See the relative weight of marks in section (a) question and section (b) question.
   b. Attempt the question carrying more marks.
   c. If partly known in section (a) and (b) judge quickly which option gives more marks over all.
   d. Never try to answer one subsection from section (a) and other subsection from section (b).
   e. Instead put more effort on the question in the same part (a) or (b).

f.   Similarly never try to answer section (a) and (b) question together. It is just waste of precious time in the examination.

g.   Make sure that if answered subsection (i) in section (a) other subsections (ii), (iii), etc are also attempted to answer.

h.   If answer is unknown to the subsections follow the Part A answering tips. However it is to be noted that if answer is unknown not to write many pages and waste the time just because it is Part B question and carries more marks. May be about a page of a related writing will serve the purpose of attempting the question.

4.   Neatly draw the diagrams and label it.

5.   Use the appropriate technical terms. Once it is used need not elaborate unless asked for in the question it self.

6.   Correctly write the equation numbers.

7.   Use standard symbols, not to adopt your own. For example the standard accepted is either dot (· Superscript) or single electron ( ) not *.

8.   Write all relevant points in comprehensive manner since it fetches more marks, each and every point is important. Need not elaborate too much on one aspect and leaving other parts untouched.

9.   After writing answer to each question leave about 5 line space and start answering the next question. It is in case after answering all questions if time permits add some more important points.

10. Try to answer as for as possible all the subsections of a particular section continuously. In case written in a different place write full question number clearly outside the margin like 12, a, (ii). It means question number 12, section 'a' subsection (ii).

QUESTION PAPER MODEL-ANALYSIS

PART A (10 x 2 MARKS = 20 MARKS)

There are 10 questions in this part. Each question carries 2 marks. There is no choice of questions. All the questions have to be answered.

Only two questions to be asked from each unit. There are five units in the syllabus. So 5 x 2 = 10 questions. In General there will not be any subsections in Part A questions.

PART B (5 x 16 MARKS = 80 MARKS)

There are five EITHER –OR questions. It is numbered as section (a) and (b). Each section can have subsections. The subsections are numbered as (i), (ii), and so on. Generally there will not be more than three subsections in each section.

Each question is dedicated to one unit. Both the EITHER-OR question has to be from same unit. No mixing of questions from other unit is allowed.

There are five units in the syllabus. So there will be five questions starting from question number 11 to 15. Each question will have section (a) and (b). It is expected to answer either section (a) or (b) not both and/or one question from section (a) and other from section (b).

STATISTICAL ANALYSIS

Since Part A only two questions from each unit predicting the question is difficult and it can be from any corner of the subject matter in the unit.

Part B question is relatively easy to predict. Since in general atleast 4 question (even if two sub-sections) it is easy to predict the questions. If concentrate each word in the unit and prepare for about 10 marks one can easily answer all part B questions.

From every unit 36 marks questions (Part A 4 marks and Part B 32 Marks) will be asked. In which one has to answer for 20 marks. So 20 x 5 units = 100 Marks. It means instead of reading messy way all units, thoroughly read the units while in worst citation. However it is expected to read all the units systematically in order to get good score in the final examination.

It is strongly recommended to study all the units in full with in-depth knowledge. In the worst case it may be poorly sufficient to thoroughly study half or about three fourth of the unit to answer all the Part B questions. Further, a word of caution is it is always a risky task to take such a risky risk.

MODEL QUESTION PAPER

Reg. No. [　|　|　|　|　|　|　|　|　|　]

B.E./B.Tech. DEGREE EXAMINATION, MONTH, YEAR.

Second Semester

ENGINEERING CHEMISTRY-II

Time : Three hours

Maximum : 100 marks

Answer ALL Questions

Part A – (10 x 2 = 20 marks)

1. What is single electrode potential and its relevance to emf?
2. Elucidate the principle behind the function of ion selective electrodes.
3. State Pilling Bedworth rule with examples.
4. What are the factors responsible for pitting corrosion?
5. Calculate the volume of air required for the complete combustion of 24 g of carbon.
6. Bring out the role of proximate analysis in fuel grading.
7. Define the terms isobar and eutectic point.
8. Give the composition and use of a non-ferrous alloy.
9. Calculate the number of vibration modes for O_3. H_2S, ClO_2 and SiO_2.
10. What are forbidden and allowed transitions in UV spectroscopy?

Part B – (5 x 16 = 80 marks)

11. a (i) Construct a cell using SHE and find out emf of an unknown electrode. (8)

   (ii) Give the salient features of electrochemical series and its significance. (8)

OR

  b. (i) Calculate the emf of the following cell at 298K. $E° = 0.635$ V.

$Zn/Zn^{2+}(0.005$ M$)//Pb^{2+}$ (0.05 M$)/Pb$ (8)

    (ii) Analyze the use of potentiometry in quantitative analysis. (8)

12. a (i) Explore the ways of corrosion protection by cathodic protection. (10)

   (ii) Briefly explain the constituent of paints and it significance. (6)

OR

  b (i) Substantiate the fact that electroplating of gold impart corrosion resistance with examples. (6)

   (ii) Electroless plating is a not an electrochemical process. Comment the statement with evidence. (4)

(iii) Explain the relevant factors responsible for the corrosion of a metal. (6)

13. a (i) What is coke and how is it produced. (10)

(ii) Calculate the calorific value of a fuel having the following composition.

C= 92 %, H = 6 %, S = 1%, N = 1%. (6)

OR

b (i) Explain the fluid bed cracking process for the manufacture of gasoline. (6)

(ii) Explain the significance of cetane number (4)

(iii) Estimation the characteristic of emission using a suitable experiment. (10)

14. (i) Explain the characteristics of biphasic equilibrium in water system. (6)

(ii) What is the principle behind thermal analysis and its relevance in the construction of phase diagram? (10)

OR

b (i) Discuss the principle of thermal annealing of steel and its impact on the properties of steel. (6)

(ii) Explain the following

Phase Rule (2)

Lever Rule (2)

Cyaniding (2)

(iii) Identify the number of phase and components in the following equilibrium. (4)

$$3Fe_2O_3 + H_2O \qquad 2Fe_3O_4.H_2O$$

15. (i) Calculate reduced mass for, H2, H-D and H-T . (6)

(ii) Analyze group vibrations in IR spectrum. (6)

(iii) Intensity of 260 nm light is reduced to 15 % when passed through 1 x 10-3 cell containing 0.05 mol dm-3 solution. Calculate molar extinction coefficient. (4)

OR

b (i) Describe a method for the estimation of alkali metals and explain the principle behind the experiment. (8)

(ii) Discuss the following terms with relevant examples (8)

Chromophore

Overtones

Dilution Factor

Nebulizer.

16. Neatly draw the diagrams and label it.

17. Use the appropriate technical terms. Once it is used need not elaborate unless asked for in the question it self.

18. Correctly write the equation numbers.

19. Use standard symbols, not to adopt your own. For example the standard accepted is either dot ($\cdot$ Superscript) or single electron ( ) not *.

20. Write all relevant points in comprehensive manner since it fetches more marks, each and every point is important. Need not elaborate too much on one aspect and leaving other parts untouched.

21. After writing answer to each question leave about 5 line space and start answering the next question. It is in case after answering all questions if time permits add some more important points.

22. Try to answer as for as possible all the subsections of a particular section continuously. In case written in a different place write full question number clearly outside the margin like 12, a, (ii). It means question number 12, section 'a' subsection (ii).